NONLINEAR EVOLUTION EQUATIONS AND POTENTIAL THEORY

NONLINEAR EVOLUTION EQUATIONS AND POTENTIAL THEORY

Edited by

JOSEF KRÁL

Mathematical Institute
of the Czechoslovak Academy of Sciences
Prague, Czechoslovakia

PLENUM PRESS · LONDON AND NEW YORK

Published in coedition with ACADEMIA,
Publishing House of the Czechoslovak
Academy of Sciences, Prague

and outside the Socialist countries by

Plenum Press, London
A division of Plenum Publishing Company Limited
4a Lower John Street, London W1R 3PD, England
(Telephone 01-437 1408)

U.S. edition published by Plenum Press, New York
A division of Plenum Publishing Corporation
227 West 17th Street, New York N.Y. 10011, U.S.A.

Lectures of a Summer School held at Podhradí near Ledeč on Sázava, Czechoslovakia, September, 1973

ISBN 0 - 306 - 30835 - 5
Library of Congress Catalog Card Number 74 - 20000

B7T

Printed in Czechoslovakia

LIST OF CONTRIBUTORS

Gottfried Anger	Sektion Mathematik, Martin Luther Universität, Halle, GDR
Viorel Barbu	Faculty of Mathematics, University of Iaşi, Romania
Haim Brezis	Department des Mathématiques, Université de Paris, France
Siegfried Dümmel	Sektion Mathematik, Technische Hochschule, Karl-Marx-Stadt, GDR
Jozef Kačur	Faculty of Natural Sciences, Komenský University, Bratislava, Czechoslovakia
Josef Král	Mathematical Institute of Czechoslovak Academy of Sciences, Prague, Czechoslovakia
Vladimír Lovicar	Mathematical Institute of Czechoslovak Academy of Sciences, Prague, Czechoslovakia
Jaroslav Lukeš	Faculty of Mathematics and Physics, Charles University, Prague, Czechoslovakia
Jiří Veselý	Faculty of Mathematics and Physics, Charles University, Prague, Czechoslovakia
Ivo Vrkoč	Mathematical Institute of Czechoslovak Academy of Sciences, Prague, Czechoslovakia

CONTENTS

PREFACE

During recent years, the Mathematical Institute of the Czechoslovak Academy of Sciences has organized summer schools devoted to non-linear functional analysis and its applications - particularly in the theory of boundary value problems for differential equations.

The main subjects of the summer school held from 24 to 29 September 1973 at Podhradí near Ledeč on Sázava were theory of nonlinear evolution equations and potential theory.

The summer school was attended by more than 60 mathematicians from Czechoslovakia and abroad. The lectures were delivered by

Gottfried Anger, Halle (GDR),
Viorel Barbu, Iași (Romania),
Haim Brezis, Paris (France),
Siegfried Dümmel, Karl-Marx-Stadt (GDR),
Jozef Kačur, Bratislava (Czechoslovakia),
Josef Král, Praha (Czechoslovakia),
S.N. Kružkov, Moskva (USSR),
Vladimír Lovicar, Praha (Czechoslovakia),
Jaroslav Lukeš, Praha (Czechoslovakia),
Jiří Veselý, Praha (Czechoslovakia),
Ivo Vrkoč, Praha (Czechoslovakia).

In the present proceedings the text of almost all lectures delivered during the school are collected.

Josef Král
Editor

September, 1974

DIRECT AND INVERSE PROBLEMS IN POTENTIAL THEORY

Gottfried Anger
Halle (GDR)

The aim of this paper is to sketch the most important direct problems (boundary value problems and initial value problems) of linear elliptic, parabolic and hyperbolic differential equations and some inverse problems corresponding to these equations. Both types of problems are divided into two classes. The first one is the class of properly posed problems, the other is the class of improperly posed problems. The Dirichlet problem for elliptic equations and parabolic equations and the Cauchy problem for hyperbolic equations are properly posed problems, the Dirichlet problem for hyperbolic equations and the inverse problem for the Laplace equation and the heat equation are improperly posed problems. There exist also inverse (improperly posed) problems concerning hyperbolic equations.

A problem is properly posed with respect to two spaces (of continuous functions), if and only if there exist certain a priori estimates in C-norms. Regarding elliptic equations of the second order we obtain estimates in the supremumnorm. For equations of higher order we obtain estimates in the supremumnorm for the function and its derivatives up to a fixed order. We sketch the results of Boboc, Mustaţa, Miranda, Agmon, Nirenberg, Schechter, Canfora, Ladyshenskaja, Solonnikov, Krasovskij and a new method developed by B.-W. Schulze.

A new method of potential theory, developed by G. Anger, G.

Wildenhain and B.-W. Schulze, occupies a large part of this paper. This method is based on continuous or m-times continuously differentiable potentials and on the notion of capacity corresponding to such potentials. Many results of potential theory are valid also for elliptic equations of higher order and elliptic systems, for instance the results on capacity, the continuity principle and the fine topology. We give a general theory for the sets of capacity zero. Our investigation is applied to the balayage-principle. We state an equivalent integral form, which is important in the investigation of inverse problems.

The last section contains a new method, developed by G. Anger for studying inverse problems of the Laplace equation and the heat equation. We consider all positive measures with the same potential outside a fixed bounded domain. This set of measures is convex and weakly compact. We characterize classes of extreme points and suggest a general construction corresponding to the measures considered. Further we consider an inverse variational principle and prove among other, that the supremum of the Dirichlet integral with respect to the measures considered is infinite.

1. EXAMPLES OF DIRECT AND INVERSE PROBLEMS

1.1 NOTATIONS

Let

$$P(x,D) = \sum_{|\alpha|\leq m} a_\alpha(x) D^\alpha$$

be a partial differential operator of order m,

$$P^*(x,D) = \sum_{|\alpha|\leq m} (-1)^{|\alpha|} D^\alpha(a_\alpha) .$$

the formal adjoint differential operator. The a_α are smooth functions in $\mathbf{R}^n$ (or in an arbitrary domain $\Omega \subset \mathbf{R}^n$), e.g. $a_\alpha \in C^{|\alpha|}(\mathbf{R}^n)$. In the case $a_\alpha = \text{const.}$ fundamental solution E exists, in the general case as distribution,

$$(E, P^*(D)\varphi) = \varphi(0) \text{ for every } \varphi \in C_0^\infty(\mathbf{R}^n) .$$

If E is locally integrable, we have the integral relation

$$\int E(x)P^*(D)\varphi(x)dx = \varphi(0) .$$

We put

$$\Phi(x,y) = E(x - y) .$$

Then we have the relation

$$(1) \quad \int \Phi(x,y)P^*(x,D)\varphi(x)dx = \varphi(y) \text{ for every } \varphi \in C_0^\infty(\mathbf{R}^n) .$$

This relation holds also for locally integrable fundamental solutions corresponding to a differential equation with variable coefficients.

Now we introduce the notion of potential. We consider a kernel

$$\Phi: \quad \mathbf{R}^n \times \mathbf{R}^n \mapsto \overline{\mathbf{R}}^1 = \{x: -\infty \leqq x \leqq +\infty\} .$$

This kernel can be a fundamental solution of (1). Let

$$(2) \quad \Phi^+ = \sup(\Phi,0), \quad \Phi^- = -\inf(\Phi,0) .$$

We have

$$\Phi = \Phi^+ - \Phi^- .$$

The adjoint kernel $\tilde{\Phi}$ is defined by

$$(3) \quad \tilde{\Phi}(x,y) = \Phi(y,x) .$$

In this paper we restrict our considerations to kernels Φ^+, Φ^- which are lower semicontinuous. In the applications this condition is satisfied. For the general case see G. ANGER [9], F.E. BROWDER [23], B.-W. SCHULZE [46] and G. WILDENHAIN [49].

Let μ be a positive measure on the space $\mathbf{R}^n$. The potential $\Phi\mu$ is defined at a point $x \in \mathbf{R}^n$ by

$$(4) \quad \Phi\mu(x) = \int \Phi^+(x,y)d\mu(y) - \int \Phi^-(x,y)d\mu(y)$$

(if $\Phi\mu(x) \neq \infty - \infty$). For an arbitrary measure $\mu = \mu^+ - \mu^-$, where μ^+ and μ^- are the corresponding positive measures, we define $\Phi\mu$ by

$$(5) \quad \Phi\mu(x) = \int \Phi(x,y)d\mu^+(y) - \int \Phi(x,y)d\mu^-(y) .$$

In the same way the potential $\Phi\mu$ is defined for a kernel

$$\Phi : \Omega \times \Omega \longmapsto \bar{\mathbf{R}}^1$$

and for a measure μ on Ω.

1.2 (DIRECT) BOUNDARY VALUE PROBLEMS

a) Ordinary differential equations

Let

$$(6) \quad P(D)u = \sum_{k=0}^{2m} a_k D^k u = 0, \quad a_{2m} = 1 .$$

The general solution is of the form

$$u = \sum_{k=1}^{2m} c_k u_k ,$$

where the u_k form a fundamental system and the c_k are arbitrary constants. The fundamental solution is of the form

$$(7) \quad E(x) = w(x)\,\Theta(x) + h(x) ,$$

where $w(0) = w'(0) = \ldots = w^{(2m-2)}(0) = 0$, $w^{(2m-1)}(0) = 1$ and Θ is the Heaviside function,

$$\Theta(x) = 0 \text{ for } x \leqq 0, \quad \Theta(x) = 1 \text{ for } x > 0 ,$$

while h is a solution of the equation (6). In the special case $u' = 0$ we have

$$(8) \qquad E(x) = \Theta(x) + c ,$$

and in the case $u'' = 0$

$$(9) \qquad E(x) = x\Theta(x) + c_1 x + c_2 .$$

In the case (6) we can write 2m conditions at the endpoints of a closed interval [a,b]. It is very interesting to consider the potential corresponding to the kernel $\Phi(x,y) = \Theta(x-y)$.

$$(10) \qquad \Phi\mu(x) = \int \Theta(x - y)d\mu(y) = \chi(x) .$$

We know, that for a positive measure μ on the real line with compact support supp μ there exists a one to one correspondence between μ and the real function χ, which is positive, monotone increasing and left-side continuous. The potential $\Phi\mu$ is therefore continuous in a point $x \in \mathbf{R}^1$, if and only if $\mu(\{x\}) = 0$. We know all continuous potentials in this case (G. ANGER [9], B.-W. SCHULZE [40], G. WILDENHAIN [48]).

b) Elliptic differential equations of second order

We take as an example the Laplace equation

$$(11) \qquad -\Delta_n u = -\sum_{i=1}^{n} \frac{\partial^2 u}{\partial x_i^2} = 0$$

and the Helmholtz equation

$$(12) \qquad -\Delta_n u - \varkappa^2 u = 0, \quad \varkappa > 0 .$$

The fundamental solution of (11) in $\mathbf{R}^1$ is of the form (9), in the plane $\mathbf{R}^2$ we have

(13) $E(x) = \frac{1}{2\pi} \log \frac{1}{|x|}$, $x \neq 0$,

and in $\mathbf{R}^n$, $n \geqq 3$,

(14) $E(x) = \frac{1}{(n-2)\,\omega_n} \quad \frac{1}{|x|^{n-2}}$, $x \neq 0$.

Fundamental solution of (12) in $\mathbf{R}^3$ is

(15) $E(x) = \frac{1}{4\pi} \frac{\cos \varkappa |x|}{|x|}$, $x \neq 0$.

For the general form in $\mathbf{R}^n$ see for instance G. ANGER [9]. For elliptic equations of second order we study the

Dirichlet problem: Determine a function $u \in C^2(\Omega) \cap C(\bar{\Omega})$, which is a solution of the equation $-\Delta_n u = 0$ in Ω (or of $-\Delta_n u - \varkappa^2 u = 0$) and which satisfies the condition $u|_{\partial\Omega} = f$, $f \in C(\partial\Omega)$, on the boundary $\partial\Omega$.

For a ball

$K_r = \{x : |x| < r\}$

we can calculate the solution u explicitly with the aid of the Poisson integral

(16) $u(x) = \int \frac{r^2 - |x|^2}{r\omega_n} \quad \frac{f(y)}{|x - y|^n} \, dO(y) = \int f(y) d\mu_x(y)$,

where μ_x is a positive measure on ∂K_r.

For elliptic equations of the second order we can consider the

Cauchy problem: In a half space $x_n \geqq 0$ determine a solution of the equation $P(x,D)u = 0$, which satisfies the conditions $u|_{x=0} = f$, $\left.\frac{\partial u}{\partial x_n}\right|_{x_n=0} = g$.

J. HADAMARD gave an important example, showing that this

problem is not properly posed (see for instance M.M. LAVRENTIEV [31] or S.G. Michlin [35]).
In the case $\Delta_n u = 0$ he considered

$$(17) \quad u_k(x) = \frac{1}{k^2} \sinh(kx_2) \sin(kx_1) .$$

We have

$$u_k \Big|_{x_2=0} = 0 , \quad \frac{\partial u}{\partial x_2}\Big|_{x_2=0} = \frac{\sin(kx_1)}{k} .$$

In the case of the Helmholtz equation there exist eigenfunctions, i.e. $u(y) = 0$ on $\partial\Omega$ does not imply $u(x) \equiv 0$ in Ω. As examples we may consider solutions of $u'' + \varkappa^2 u = 0$ in $\mathbf{R}^1$. If for a domain $\Omega \subset \mathbf{R}^n$ eigenfunctions do not exist, we can write the solution in the form

$$(18) \quad u(x) = \int f(y) d\mu_{x,\varkappa}(y) ,$$

where the measure is of the form

$$\mu_{x,\varkappa} = \mu^+_{x,\varkappa} - \mu^-_{x,\varkappa} .$$

(see for instance G. ALBINUS [4] and G. ANGER and B.-W. SCHÜLZE [16]).

c) Parabolic equations of the second order

We consider as an example the heat equation

$$(19) \quad \Delta_{n-1} u - \frac{\partial u}{\partial x_n} = 0 .$$

The fundamental solution has the form

$$(20) \quad E(x) = \left(\frac{1}{2\sqrt{\pi x_n}}\right)^{n-1} \exp\left(-\sum_{i=1}^{n-1} x_i^2 / 4x_n\right) \text{ for } x_n > 0 ,$$

$$E(x) = 0 \text{ for } x_n = 0 .$$

For the heat equation we can also pose the Dirichlet problem. Let

$$\partial\Omega_x = \{z \in \partial\Omega,\ z_n \leqq x_n\}.$$

Then the solution of the Dirichlet problem (corresponding to a special domain $\Omega_x = \{z \in \Omega,\ z_n < x_n\}$) is of the form

$$(21) \qquad u(x) = \int f(y)\,d\mu_x(y)\ ,\quad \mu_x \geqq 0\ ,\quad \operatorname{supp}\mu_x \subset \partial\Omega_x\ .$$

d) Elliptic equations of higher order

We take as an example the polyharmonic equation

$$(22) \qquad \Delta_n^p u = \Delta_n(\Delta_n^{p-1}u) = 0\ .$$

In the space $\mathbf{R}^1$ we have the equation u'''' = 0. The fundamental solution in $\mathbf{R}^1$ has the form $\frac{x^3}{3!}\Theta(x)$; in $\mathbf{R}^n$, $n \geqq 2$, we have

$$(23) \qquad E(x) = |x|^{2p-n}(A_{p,n}\log|x| + B_{p,n})\ ,$$

where $A_{p,n}$ and $B_{p,n}$ are suitable constants. It is $A_{p,n} = 0$ in the case $2p - n < 0$ or if $2p - n \geqq 0$ and n is odd. It is $B_{p,n} = 0$ in the case $2p - n \geqq 0$ and n even.

Let us formulate the Dirichlet problem in the case $p = 2$ for a smooth domain Ω:

Dirichlet problem: Find a function $u \in C^4(\Omega) \cap C^1(\bar\Omega)$, which is a solution of the equation $\Delta_n^2 u = 0$ in Ω and which assumes on $\partial\Omega$ the prescribed values $u\big|_{\partial\Omega} = f$, $\frac{\partial u}{\partial n}\Big|_{\partial\Omega} = g$.

For a ball we know a Poisson formula, $n = 2$,

$$(24) \qquad u(x) = \frac{1}{2\pi r}(|x|^2 - r^2)\left[\frac{1}{2}\int_0^{2\pi}\frac{-g(y)\,d\varphi(y)}{|x-y|^2} + \int_0^{2\pi}\frac{f(y)(r - |x|\cos(\varphi-\Theta))\,d\varphi(y)}{|x-y|^2}\right]$$

where $y = (r,\varphi)$, $x = (|x|,\Theta)$.

We can also formulate the Dirichlet problem with the boundary conditions

$$u\Big|_{\partial\Omega} = f, \quad \frac{\partial u}{\partial x_i}\Big|_{\partial\Omega} = g_i ,$$

where f and g_i are the restrictions of a function F and its first derivatives, defined in the space $\mathbf{R}^n$, to $\partial\Omega$. In this case it follows from (24) for a ball K_r that

$$(25) \qquad u(x) = \int f(y)\,d\mu_x(y) + \sum_{i=1}^{n} \int g_i(y)\,d\mu_{x,i}(y) ,$$

where μ_x and $\mu_{x,i}$ are measures on ∂K_r.

For an elliptic equation we can formulate general (local) boundary conditions (see e.g. G. ANGER, B.-W. SCHULZE and G. WILDENHAIN [15])

$$(26) \qquad Q_\beta u\Big|_{\partial\Omega} = \sum_{|\alpha| \leqq m_i} a_{\alpha\beta}(y) D^\alpha u\Big|_{\partial\Omega} = g_\beta , \quad |\beta| \leqq p-1 .$$

e) Other equations

We take as an example the equation

$$(27) \qquad \frac{\partial^n u}{\partial x_1 \ldots \partial x_n} = 0 .$$

In the case n = 2 the equation is a hyperbolic equation, the so called d'Alembert equation. The fundamental solution is of the form

$$(28) \qquad E(x) = \Theta(x_1) \ldots \Theta(x_n) .$$

The solution of the Cauchy problem has in the case n = 2 the form

$$u(x) = \frac{f(x_1+x_2) + f(x_1-x_2)}{2} + \frac{1}{2} \int_{x_1-x_2}^{x_1+x_2} g(\xi)\,d\xi .$$

For this eguation we can also study a Dirichlet problem for special domains (see e.g. B.-W. SCHULZE [40], [41]).

1.3 SOME INVERSE PROBLEMS

First let us consider some inverse problems in $\mathbf{R}^1$ for the ordinary differential equation (2). We consider the potentials (10) for the equation $u' = 0$. For a positive measure ν with compact support supp ν we have

$$\Phi\nu(x) = \int \Theta(x - y)d\nu(y) = \chi(x) .$$

We consider the set $\mathfrak{M}_{\delta_{x^o}}$ of all measures $\nu \geqq 0$ with supp $\nu \leqq x^o$ and

$$(29) \quad \chi(x) = \Phi\nu(x) = \Phi\delta_{x^o}(x) \text{ for } x > x^o .$$

Let g be a monotone increasing continuous function with $g(x) = \Phi\delta_{x^o}(x)$ for all $x > x^o$. We can also consider the subset $\mathfrak{M}^g_{\delta_{x^o}}$ consisting of all measures $\nu \in \mathfrak{M}_{\delta_{x^o}}$ with

$$(30) \quad \Phi\nu(x) \leqq g(x) \text{ for all } x \in \mathbf{R}^1 .$$

In the case $u'' = 0$ we consider the fundamental solution (9). Let $\mu = s\delta_a + (1 - s)\delta_b$, $0 \leqq s \leqq 1$, and $\Omega = (a,b)$. Then we consider the set $\mathfrak{M}_\mu$ of all measures ν with supp $\nu \subset \bar{\Omega}$ and

$$(31) \quad \Phi\nu(x) = \Phi\mu(x) \text{ for all } x \in \mathbf{R}\setminus\bar{\Omega} .$$

The problem (31) can be posed for the equation (11), (12), (19) and (22). In this case Φ is the fundamental solution of the differential equation considered and $[a,b]$ is replaced by the closure of a bounded domain $\Omega \subset \mathbf{R}^n$. We can subject ν to further conditions, for instance to be of the form $d\nu(y) = \varrho(y)dy$ or $d\nu(y) = \sigma(y)dO(y)$. In the case of the Laplace equation see A.I. PRILEPKO [37] and for other inverse problems see for instance M.M. LAVRENTIEV [31], M.M. LAVRENTIEV, V.O. ROMANOV and VASILIEV [32], V.O. ROMANOV [38], A.N. TYCHONOV, V.K. IVANOV and M.M. LAVRENTIEV [47] and [53].

2. PROPERLY POSED PROBLEMS

This type of problems was introduced by J. HADAMARD. Let B_1, B_2 be complete metric spaces

$$A : B_1 \longmapsto B_2$$

a mapping. The problem

$$(1) \quad Au = g$$

is properly posed, if the following conditions are satisfied:

(2)
1. The solution of (1) exists for any $g \in B_2$,
2. The solution of (1) is unique in B_1,
3. The solution of (1) depends continuously on the right-hand side g.

The problem (1) is improperly posed, if any of these three conditions is not fulfilled. From 1, 2, 3 it follows that the inverse mapping A^{-1} exists and is continuous. In the special case of the Dirichlet problem for the Laplace equation (11) we take for B_1 the set of all functions $u \in C(\bar{\Omega})$, which are solutions of the equation (11), and for B_2 the set of all restrictions $Au = \text{rest}_{\partial\Omega}\, u$ on $\partial\Omega$. We have $B_2 \subset C(\partial\Omega)$ and $B_2 = C(\partial\Omega)$ if and only if the boundary points $y \in \partial\Omega$ are regular (G. ANGER [11], [13]). If B_1 and B_2 are Banach spaces and A is linear, it follows from $u = A^{-1}g$ that

$$(3) \quad \|u\|_1 = M\|g\|_2 = M\|\text{rest}_{\partial\Omega}\, u\|_2 .$$

In our special case we have

$$(4) \quad \sup_{x \in \bar{\Omega}} |u(x)| \leqq M \sup_{y \in \partial\Omega} |u(y)| , \quad M = 1 .$$

This inequality can be proved for the Laplace equation by the strong maximum-minimum-principle

$$(5) \quad \inf_{y \in \partial\Omega} u(y) \leqq u(x) \leqq \sup_{y \in \partial\Omega} u(y) .$$

The inequality (4) holds also for the equation (12) under the condition

(6) if $u(y) \equiv 0$ on $\partial\Omega$, then we have $u(x) \equiv 0$ in Ω.

This inequality was proved by N. BOBOC and P. MUSTAŢĂ [19]. The case of eigenfunctions was studied by G. ALBINUS, N. BOBOC and P. MUSTAŢĂ [5].

For the Laplace equation we have a largely developed theory of the Dirichlet problem and related problems. The reason is the existence of the maximum-minimum-principle (5). This inequality implies the following integral representation for the solution u of the Dirichlet problem

$$(7) \quad u(x) = \int f(y)d\mu_x(y), \quad \mu_x \geqq 0, \quad \operatorname{supp} \mu_x \subset \partial\Omega .$$

At all regular boundary points $y \in \partial\Omega$ (i.e. $G(x,y) = 0$) we have $u(x) \to f(y)$. This theory is developed in the books and papers of H. BAUER [18], M. BRELOT [20], [21], [22], C. CONSTANTINESCU and A. CORNEA [27].

In the case of the Helmholtz equation (12) we can prove under the assumption (6) only the inequality (4), from which follows

$$(8) \quad u(x) = \int f(y)d\mu_{x,\varkappa}(y), \quad \mu_{x,\varkappa} = \mu^+_{x,\varkappa} - \mu^-_{x,\varkappa},$$

$$\operatorname{supp} \mu_{x,\varkappa} \subset \partial\Omega .$$

It was proved by G. ALBINUS [4], that $\mu_{x,\varkappa}$ is absolutely continuous with respect to the harmonic measure $\mu_x = \mu^0_x$. The regular boundary points are the same with respect to the Laplace equation and the Helmholtz equation. For any regular point $y \in \partial\Omega$ we have

$$(9) \quad \mu^+_{x,\varkappa} \to \delta_y, \quad \mu^-_{x,\varkappa} \to 0$$

in the sense of the weak topology and $\|\mu^-_{x,\varkappa}\| \to 0$ (see G. ANGER and B.-W. SCHULZE [16]).

For the heat equation we have also a strong maximum-minimum-principle of the form (5), where $\partial\Omega$ must be replaced by $\partial\Omega_x$. Therefore a potential theory exists also for the heat equation (see for instance H. BAUER [18], C. CONSTANTINESCU and A. CORNEA

[27]). However, there are differences between the behaviour of the solutions of the Laplace equation and the solutions of the heat equations (see e.g. J. LUKEŠ [33]).

Now we consider the polyharmonic equation (22). Let $C^p(\Omega)$ be the set of functions $u \in C(\Omega)$ p-times continuously differentiable up to the boundary $\partial\Omega$. The solution u of the Dirichlet problem of the biharmonic equation for a ball has the form (see formula (24) of Section 1)

$$u(x) = \sum_{|\alpha| \leqq 1} \int D\, u(y) d\mu_{x,\alpha}(y) .$$

It is very difficult to prove such equalities for elliptic equations of higher order. B.-W. SCHULZE has developed a new method for proving such equalities. He uses results of MIRANDA [36], AGMON [1], AGMON-DOUGLIS-NIRENBERG [2], SCHECHTER [39] and others in his theory. See Section 4 of our paper.

Improperly posed problems are the Cauchy problem for elliptic equations (see example (17) of Section 1, F.E. BROWDER [23] and M. M. LAVRENTIEV [31]), the Dirichlet problem for hyperbolic equations (see e.g. B.-W. SCHULZE [40], [41]) and the inverse problems in Section 1.3.

3. POTENTIALS, CONTINUOUS POTENTIALS, CAPACITY

Let Φ be a locally integrable fundamental solution of a linear differential equation. For such a kernel we can define potentials $\Phi\mu$. In our theory, continuous potentials are important. With the aid of such potentials we can introduce the notion of capacity with respect to a kernel Φ.

Let us introduce the following sets (see G. ANGER [6] - [10])

(1) $\mathfrak{F}^+(\Phi) = \{\lambda \geqq 0,$ supp λ compact, $\Phi^+\lambda$ and $\Phi^-\lambda$ continuous$\}$,

(2) $\mathfrak{F}(\Phi) = \{\lambda = \lambda_1 - \lambda_2,\ \lambda_1, \lambda_2 \in \mathfrak{F}^+(\Phi)\}$.

We know all continuous potentials for the fundamental solutions (8) and (28) of Section 1. In the first case these measures are all measures λ with $\lambda(\{x\}) = 0$ for all $x \in \mathbf{R}^1$ (G. ANGER [9], G. WILDENHAIN [48]). In the other case we have to use planes (B.-W. SCHULZE [40], [41]).

Further we consider the sets

$$(3) \quad \mathfrak{E}^+(\Phi) = \{\mu \geqq 0, \int \Phi^+ \mu d\mu < \infty, \int \Phi^- \mu d\mu < \infty\},$$

$$(4) \quad \mathfrak{E}(\Phi) = \{\mu = \mu_1 - \mu_2, \mu_1, \mu_2 \in \mathfrak{E}^+(\Phi)\}.$$

The restriction λ_K of a measure λ to a compact set $K \subset \mathbf{R}^n$ is defined by

$$\lambda_K(f) = \int f(y) c_K(y) d\lambda(y), \quad f \in C_o(\mathbf{R}^n).$$

Here c_K is the characteristic function of K. For an arbitrary compact set K we put

$$(5) \quad \begin{aligned} \mathfrak{F}^+_K(\Phi) &= \{\lambda \in \mathfrak{F}^+(\Phi), \ \operatorname{supp} \lambda \subset K\}, \\ \mathfrak{E}^+_K(\Phi) &= \{\mu \in \mathfrak{E}^+(\Phi), \ \operatorname{supp} \mu \subset K\}. \end{aligned}$$

In the case of the Newton kernel (14) of Section 1 we write $\mathfrak{F} = \mathfrak{F}(\Phi)$. In all further considerations Φ^+ and Φ^- are lower semicontinuous. In the applications this is always the case. The semicontinuity of $\Phi^+\mu$ follows for $\mu \geqq 0$ from the semicontinuity of Φ^+ (see e.g. M. BRELOT [20]). The following theorem is a consequence of the semicontinuity of Φ^+ and Φ^-.

Theorem 1: *For every $\lambda \in \mathfrak{F}^+(\Phi)$ and for every compact set $K \subset \mathbf{R}^n$ we have $\lambda_K \in \mathfrak{F}^+(\Phi)$.*

In our considerations we use only sets of capacity zero. For sets of positive capacity see G. CHOQUET [25], M. BRELOT [20] and V.G. MAZJA and V.P. CHAVIN [34]. There are also other definitions of capacity, which are equivalent to ours in some cases (see E.M. LANDIS [30], G. WILDENHAIN [50]).

Definition 1: *A set $B \subset \mathbf{R}^n$ is said to be of (inner) Φ-capacity* *zero (in symbols* $\operatorname{cap}_\Phi B = 0$), *if there exists no measure* $\lambda \in \mathfrak{F}^+(\Phi)$, $\lambda \neq 0$, *with* $\operatorname{supp} \lambda \subset B$.

If Φ is not continuous for $x \neq y$, the set $\{x^0\}$ is of capacity zero. Let $\mathfrak{B}$ be the system of all Borel sets $B \subset \mathbf{R}^n$,

$$(7) \quad J_\mu = \{B \in \mathfrak{B},\ \mu(B) = 0\} .$$

We denote by J_Φ the set of all Borel sets with Φ-capacity zero.

Theorem 2: *A set* $B \in \mathfrak{B}$ *is of* Φ*-capacity zero if and only if* $\lambda(B) = 0$ *for every* $\lambda \in \mathfrak{F}^+(\Phi)$.

Theorem 3: *Let* $K \subset \mathbf{R}^n$ *be a compact set. The following conditions are equivalent*

a) $\tilde{\Phi}\mu_1(y) = \tilde{\Phi}\mu_2(y)$ *on* K, *except a set of* Φ*-capacity zero,*

b) $\int \tilde{\Phi}\mu_1 d\lambda = \int \tilde{\Phi}\mu_2 d\lambda$ *for every* $\lambda \in \mathfrak{F}_K^+(\Phi)$,

c) $\int \Phi\lambda\, d\mu_1 = \int \Phi\lambda\, d\mu_2$ *for every* $\lambda \in \mathfrak{F}_K^+(\Phi)$.

Let

$$C_\infty(\mathbf{R}^n) = \{f \in C(\mathbf{R}^n),\ f(x) \to 0 \text{ for } |x| \to \infty\} .$$

In the sequel we use the Newton kernel Φ. For every $\lambda \in \mathfrak{F}^+$ we have $\Phi\lambda \in C_\infty(\mathbf{R}^n)$. Let us calculate the following relation, $\varphi \in C_0(\mathbf{R}^n)$:

$$(8) \quad L_\mu^0(\varphi) = \int \Phi\mu(x)\varphi(x)dx = \int\left(\int \Phi(x,y)d\mu(y)\right)\varphi(x)dx = \\ = \int\left(\int \Phi(x,y)\varphi(x)dx\right)d\mu(y) = \int \Phi\lambda_\varphi\, d\mu = \mu(\Phi\lambda_\varphi) ,$$

where $d\lambda_\varphi(x) = \varphi(x)dx$. It is $\lambda_\varphi \in \mathfrak{F}^+$ (G. ANGER [12]). To calculate $L_\mu(\varphi)$ we must know the values of $\Phi\mu$ on the support of φ. $\operatorname{supp}\varphi$ is the closure of an open set, because $\varphi \in C_0(\mathbf{R}^n)$. We have further for $\varphi = -\Delta_n\psi$, $\psi \in C_0(\mathbf{R}^n)$,

$$L_\mu(\varphi) = \int -\Phi\lambda_{\Delta_n\varphi}\, d\mu = \int \psi\, d\mu .$$

From

$$L_{\mu_1}(\varphi) = L_{\mu_2}(\varphi) \quad \text{for every } \varphi \in C_0(\mathbf{R}^n)$$

it follows

$$\int \psi d\mu_1 = \int \psi d\mu_2 \quad \text{for every } \psi \in C_o(\mathbf{R}^n) .$$

Because of the density of $C_o(\mathbf{R}^n)$ in $C_\infty(\mathbf{R}^n)$, we have $\mu_1 = \mu_2$.

Now we take a measure $\lambda \in \mathfrak{F}_K$, $\lambda \neq 0$. Then we get for all measures μ with supp $\mu \subset K$

$$(9) \quad \lambda(\Phi\mu) = \int \Phi\mu \, d\lambda = \int \Phi\lambda \, d\mu = \mu(\Phi\lambda) .$$

If K is a set of positive Φ-capacity, we can calculate the value $\mu(\Phi\lambda)$ from the value of $\Phi\mu$ on K.

Let

$$(10) \quad D(\Phi) = \{ g = \Phi\lambda \, , \, \lambda \in \mathfrak{F}(\Phi) \} ,$$

$$(11) \quad D(K;\Phi) = \{ f \in C(K), \, f(x) = \Phi\lambda(x), \, \lambda \in \mathfrak{F}_K(\Phi) \} .$$

The case

$$(12) \quad \overline{D}(K;\Phi) = C(K)$$

is of particular importance.

Theorem 4: $\overline{D}(K;\Phi) = C(K)$ *if and only if* K *consists only of regular boundary points (stable points in the sense of Keldysh) of the domain* $\mathbf{R}^n \setminus K$.

A proof of this theorem was given by G. ANGER [11], [13].

We know completely the sets of capacity zero corresponding to the kernels (8) and (28) of Section 1. G. WILDENHAIN [50] proved that a set $B \subset \mathbf{R}^1$ is of Φ-capacity zero with respect to the kernel (8), if and only if every compact subset $K \subset B$ is at most countable. B.-W. SCHULZE [40] proved an analoguous theorem for the kernel (28).

The kernels of the polyharmonic equation produce continuous potentials in the case $2p > n$. In this case the potentials $\Phi\mu$ are continuous. In our considerations we can demand, that the potentials should have continuous partials up to the order $2p - 1$. In this way we can define $2p - 1$ different notions of capacity (see G. ANGER [9]). In the investigations of the Dirichlet problem for the polyharmonic equation we use potentials with continuous derivatives of order 2p - 2 (see G. WILDENHAIN [49], B.-W. SCHULZE [46]). The sets of capacity zero formed with these potentials are the same as the sets of capacity zero in Wiener's sense for the Laplace equation. These ideas were transformed to be suitable to systems of differential equations (B.-W. SCHULZE [46]). For other possibilities of defining a capacity for equations of higher order see e.g. V.G. MAZJA and V.P. CHAVIN [34].

Now our definitions of capacity. We take as an example the kernel of the polyharmonic equation. (For the general case see G. ANGER [9] and B.-W. SCHULZE [46].) Let

$$
\begin{aligned}
&\Phi_\alpha^\beta(x,y) = D_x^\alpha D_y^\beta \, \Phi(x,y) \text{ for } x \neq y, \\
(13)\qquad &\Phi_\alpha^\beta(x,x) = D_x^\alpha D_y^\beta \, \Phi(x,x), \\
&\Phi_\alpha^\beta(x,x) = 0, \text{ if the derivative does not exist.}
\end{aligned}
$$

We define

$$
\begin{aligned}
(14)\qquad &[\Phi_\alpha^\beta(x,y)]^+ = \sup\,(\Phi_\alpha^\beta(x,y),0)\ , \\
&[\Phi_\alpha^\beta(x,y)]^- = -\inf\,(\Phi_\alpha^\beta(x,y),0)\ .
\end{aligned}
$$

The kernels (14) corresponding to the kernel of the polyharmonic equation are lower semicontinuous. In the general case we demand this property. Every distribution L of order m can be represented by a system

$$\lambda = (\lambda^\beta)_{|\beta| \leqq m} \in X\mathfrak{M}(\mathbf{R}^n)$$

of measures λ^β. This representation is not unique in general. We have

$$(15)\qquad L(f) = \sum_{|\beta| \leqq m} \int D^\beta f \, d\lambda^\beta, \quad f \in C^m(\mathbf{R}^n)\ .$$

We define the potential relative to $\lambda = (\lambda^\beta)_{|\beta| \leqq m}$ by

$$(16)\qquad \Phi\lambda(x) = \sum_{|\beta| \leqq m} \int \Phi^\beta(x,y) \, d\lambda^\beta(y)\ .$$

Often we use the potentials

$$(17)\qquad \Phi_\alpha\lambda(x) = \sum_{|\beta|\leqq m} \int \Phi_\alpha^\beta(x,y)\,d\lambda^\beta(y)\ ,\quad |\alpha| \leqq m\ .$$

The adjoint potential $\Phi^*\varkappa$ relative $\varkappa = (\mu_\alpha)_{|\alpha|\leqq m}$ is defined by

$$(18)\qquad \Phi^*\varkappa(y) = \sum_{|\alpha|\leqq m} \int \Phi_\alpha(x,a)\,d\mu_\alpha(x)\ .$$

The necessity of defining potentials (16) - (18) arises from the study of the Dirichlet problem for elliptic differential equations of higher order, see e.g. formula (25) of Section 1. Let Φ be the polyharmonic kernel and

(19) $\mathfrak{F}^+(\Phi\,;2p-2)$ the set of all positive measures λ with compact support and continuous potentials $[\Phi^\beta]^{\pm}\lambda\ ,\ |\beta| \leqq 2p-2$.

Further let

$$(20)\qquad \mathfrak{F}_K^+(\Phi\,;2p-2) = \{\lambda \in \mathfrak{F}^+(\Phi\,;2p-2),\ \operatorname{supp}\lambda \subset K\}\ .$$

Definition 2: *A set* $B \subset \mathbf{R}^n$ *is said to be of* Φ^{2p-2} *- capacity zero (in symbols* $\operatorname{cap}_\Phi^{2p-2} B = 0$*), if there exists no measure* $\lambda \in \mathfrak{F}(\Phi\,;2p-2)$, $\lambda \neq 0$, *with* $\operatorname{supp}\lambda \subset B$.

For the above sets of capacity zero the same theorems hold as for the Laplace equation. Let

$$\mu = (\mu_\alpha)_{|\alpha|\leqq p-1},\quad \lambda = (\lambda^\beta)_{|\beta|\leqq p-1},\quad \lambda^\beta \in \mathfrak{F}(\Phi\,;p-1)\ .$$

Then we have (for the kernel of the polyharmonic equation) the relations (see G. ANGER [9], B.-W. SCHULZE [46], G. WILDENHAIN [49])

$$(20)\qquad \begin{aligned} \mu(\Phi\lambda) &= \sum_{|\alpha|\leqq p-1} \int D_x^\alpha(\Phi\lambda(x))\,d\mu_\alpha(x) \\ &= \sum_{|\alpha|\leqq p-1} \int D_x^\alpha\Big(\sum_{|\beta|\leqq p-1} \int D_y^\beta \Phi(x,y)\,d\lambda^\beta(y)\Big)\,d\mu_\alpha(x) \\ &= \sum_{|\beta|\leqq p-1} \int D_y^\beta\Big(\sum_{|\alpha|\leqq p-1} \int D_x^\alpha \Phi(x,y)\,d\mu_\alpha(x)\Big)\,d\lambda^\beta(y)) \\ &= \lambda(\Phi\mu)\ . \end{aligned}$$

There exists a general form of defining the sets of capacity zero (B.-W. SCHULZE [42], [46]). Let $\mathfrak{P}$ be a set of elements and $\mathfrak{C}$ a set of nonnegatives measures

$$T : \mathfrak{C} \longmapsto \mathfrak{P}$$

a mapping. If $A \subseteqq \mathfrak{P}$ is a subset with $TO \in A$, we have the following

Definition 3: *A set* $B \in \mathfrak{B}$ *is of the* $A(T)$ *- capacity zero, if and only if there exists no measure* $\lambda \neq 0$, $\lambda \in \mathfrak{C}$, *with* $\operatorname{supp} \lambda \subset B$, $T\lambda \in A$.

In our special cases we have

1. Φ Newton kernel: $A = C(\mathbf{R}^n)$, $T\lambda(x) = \int \Phi(x,y) d\ (y)$, $\mathfrak{C} = \mathfrak{F}^+(\Phi)$,
2. Φ kernel of the polyharmonic equation: $A = C^{2p-2}(\mathbf{R}^n)$, $T\lambda(x) = \int \Phi(x,y) d\lambda(y)$, $\mathfrak{C} = \mathfrak{F}\ (\Phi ;2p-2)$.

In the first case we can take also $\mathfrak{C} = \mathfrak{E}^+(\Phi)$ (G. ANGER [9]) and in the other $\mathfrak{C} = \mathfrak{E}^+(\Phi ;2p-2)$ (B.-W. SCHULZE [46]), where

(21) $\mathfrak{E}^+(\Phi ;2p-2)$ is the set of all positive measures μ with

$$\int \left(\int [\Phi_\alpha(x,y)]^{\pm}\, d\mu(y) \right) d\mu(x) < +\infty \quad \text{for all } \alpha \text{ with } |\alpha| \leqq 2p-2.$$

To prove these facts, we use the continuity principle for the Newton kernel. It asserts: if the potential $\Phi\mu$ of a positive measure with compact supp μ is continuous on supp μ with respect to the induced topology, then it is continuous on the whole space $\mathbf{R}^n$. With the aid of the continuity principle and the fact that $\Phi\mu$ is lower semicontinuous we can prove that every $\mu \in \mathfrak{E}^+(\Phi)$ can be represented in the form

$$\mu(f) = \lim_{k \to \infty} \lambda_k(f), \quad \lambda_k \leqq \lambda_{k+1}, \quad \lambda_k \in \mathfrak{F}^+(\Phi) .$$

If $\lambda(B) = 0$ for all $\lambda \in \mathfrak{F}^+(\Phi)$, then also $\mu(B) = 0$ for every $\mu \in \mathfrak{E}^+(\Phi)$. This means, that $\mathfrak{C} = \mathfrak{F}^+(\Phi)$ and $'\mathfrak{C} = \mathfrak{E}^+(\Phi)$ define the same sets of capacity zero. This principle is not known for the heat equation. Any set B, whose points lie in the plane

x_n = const., is always of capacity zero (G. ANGER [9]). The notion of capacity for the heat equation was also introduced by E.M. LANDIS [30].

We have a continuity principle also for the kernel of the polyharmonic equation. We numerate the kernels

$$[\Phi_\alpha]^{\pm}_{|\alpha| \leqq 2p-2} \text{ by } H_1, \ldots, H_N$$

and put

$$(22) \quad H(x,y) = \sup_{1 \leqq k \leqq N} H_k(x,y) .$$

Then we have $H_k \leqq H$. We can obtain directly for the kernel of the polyharmonic equation the inequalities ($n \leqq 3$)

$$(23) \quad c_1 \, \Phi(x,y) \leqq H(x,y) \leqq c_2 \, \Phi(x,y) ,$$

where Φ is the Newton kernel. By means of (22), (23) and the notion of a uniformly convergent integral we can prove the continuity principle for the polyharmonic kernel (B.-W. SCHULZE [46]). Perhaps it is possible to prove a continuity principle for all kernels which are fundamental solutions of elliptic equations of higher order.

Theorem 5: *The kernel (22) defined with the aid of the kernel of the polyharmonic equation satisfies the continuity principle. This condition is equivalent the following condition: For every positive measure μ with compact* K = supp μ *the continuity of* $\Phi^{\pm}_{\alpha}\mu$, $|\alpha| \leqq 2p-2$, *follows from the continuity of* $\Phi^{\pm}_{\alpha}\mu$, $|\alpha| \leqq 2p-2$, *with respect to the induced topology on* K.

We have also the following theorem, proved by G. WILDENHAIN [49], [50], and in a more general form by G. ALBINUS [3].

Theorem 6: *The kernel Φ of the polyharmonic equation and the Newton kernel Φ_N satisfy $\mathfrak{F}^+(\Phi_N) = \mathfrak{F}^+(\Phi, 2p-2)$. Therefore the sets of capacity zero are the same for both equations.*

4. NEW RESULTS ON A PRIORI ESTIMATES IN C-NORMS

In Section 2 we have considered a priori estimates for elliptic equations of the second order. In this section we shall sketch some new results by B.-W. SCHULZE [45], [46]. It is an open problem to prove such estimates in C-norms for general boundary value problems and for the Dirichlet problem for more general domains.[1)]

Let

(1) $$P(x,D) = \sum_{|\alpha|\leqq 2m} a^{\alpha}(x)D^{\alpha}$$

be an elliptic operator of order 2m with sufficiently smooth coefficients in $\mathbf{R}^n$. We consider solutions

$$u \in C^{2m}(\Omega) \cap C^{m-1}(\bar{\Omega})$$

in a bounded domain Ω ($\partial\Omega$ of class C^{∞}) of the Dirichlet problem

$$P(x,D)u = 0 \quad \text{in } \Omega$$

$$\left(\frac{\partial}{\partial n}\right)^{j-1} u\Big|_{\partial\Omega} = g_j \quad \text{on } \partial\Omega .$$

The boundary function g_j is assumed to be in $C^{m-j}(\partial\Omega)$. In recent years methods of potential theoretic character have been developed for this problem, which are analogous to certain methods for this problem in the case m = 1.

Our aim is to prove inequalities of the form

(3) $$\sum_{|\alpha|\leqq m-1} \sup_{x\in\Omega}|D^{\alpha}u(x)| \leqq M \sum_{|\alpha|\leqq m-1} \sup_{y\in\partial\Omega}|D^{\alpha}u(y)|$$

for all

$$u \quad C^{2m}(\Omega) \cap C^{m-1}(\bar{\Omega}) ,$$

which solve the problem (2). In our investigations the uniqueness of the solution is required. Otherwise we have to require that u

---

1) In October 1973 Professor V.G. Mazja informed me that he had proved such a priori estimates for domains with corners. For estimates in C-norms see also J. Nečas, Les Méthodes directes en théorie des équations elliptiques, Prague 1967.

is in a topological complement of the finite dimensional null space or we have to add the term $\|u\|_{L^1(\Omega)}$ on the right-side.

There are various methods to prove (3). We will sketch one possible way due to C. MIRANDA [36], S. AGMON [1], and a contribution of M. SCHECHTER [39], who improved and simplified the proofs of L^p-regularity for solutions of elliptic equations. This method can be extended to classes of elliptic systems (strongly elliptic systems - see A. CANFORA [24] and B.-W. SCHULZE [45], [46] - and systems, which are Douglis-Nirenberg-elliptic of a type similar to the linearized Navier-Stokes-system [45], [46]). The method consists in the construction of a function z, which has the same boundary values $g = (g_1,\dots,g_m)$ as u and depends on g continuously in the sense of (3), such that

$$(4) \qquad \sum_{|\alpha| \leqq m-1} \sup_{x \in \Omega} D^{\alpha} z(x) \leqq M\, G$$

holds, where

$$G = \sum_{j=1}^{m} \sup_{y \in \partial\Omega} |g_j(y)|$$

is the norm of g in the product space

$$\bigtimes_{j=1}^{m} C^{m-j}(\partial\Omega) \,.$$

This function has the additional property

$$(5) \qquad |(u - z, P^{*}w)| \leqq M\, G \; \|w\|_{m,p'} \quad (p' > 1)$$

for all $w \in C^{\infty}(\bar{\Omega})$ with

$$\left(\frac{\partial}{\partial n}\right)^{j-1} w \,\Big|_{\partial\Omega} = 0, \quad j = 1,\dots,m \,.$$

Now we have from well known theorems (S. AGMON [1], M. SCHECHTER [39])

$$(6) \qquad \|u - z\|_{m,p} \leqq M(G + \|u - z\|_{0,p}), \quad p = p'/p' - 1 \,,$$

and $\|u - z\|_{0,p}$ can be omitted since uniqueness is assumed. This step is one of the main difficulties for elliptic systems. The methods of M. Schechter which use interpolation of Sobolev-spaces in bounded domains are essential. From (6) we obtain in virtue of Sobolev's imbedding theorem

$$(7) \quad \|u - z\|_{C^{m-1}(\bar{\Omega})} \leqq M\, G$$

($p > 1$ sufficiently large). Hence from $u = (u - z) + z$ and (7) we have

$$(8) \quad \|u\|_{C^{m-1}(\bar{\Omega})} \leqq \|u - z\|_{C^{m-1}(\bar{\Omega})} + \|z\|_{C^{m-1}(\bar{\Omega})} =$$

$$= M(G + \|z\|_{C^{m-1}(\bar{\Omega})}) .$$

Now (4) and (8) yield the desired estimate (3).

Because of the use of Poisson-kernels in the half space and diffeomorphisms of half spheres onto boundary patches of $\partial\Omega$ the smoothness of $\partial\Omega$ cannot be essentially weakened. It is an open problem whether or not an analogous inequality in the formulation (3) holds for arbitrary domains, in which uniqueness is assumed. The method of Ju.P. KRASOVSKIJ [32] also uses considerations in the half space, but (3) follows then from estimates of the singularities of the derivatives of Green's function and the Poisson kernels. In all these cases the proofs are very technical.

A simple example for systems is the Dirichlet problem for the linearized Navier-Stokes system

$$\Delta u_i - \frac{\partial}{\partial x_i} u_4 = 0 \ (i=1,2,3), \quad \sum_{i=1}^{3} \frac{\partial}{\partial x_i} u_i = 0 \quad \text{in } \Omega$$

$$u_i\big|_{\partial\Omega} = g_i \in C(\partial\Omega), \ i = 1,2,3, \quad x = (x_1,x_2,x_3) .$$

Then the solutions

$$u = (u_1,\ldots,u_4) \in [C^2(\Omega)]^3 \times C^1(\Omega) \cap [C(\bar{\Omega})]^4$$

satisfy the estimate

$$\sum_{i=1}^{3} \sup_{x\in\Omega} |u_i(x)| \leqq M \sum_{i=1}^{3} \sup_{y\in\partial\Omega} |u_i(y)| .$$

This special result was proved earlier by S.L. SOLONNIKOV and O.A. LADYZHENSKAJA and in the last year by B.-W. SCHULZE [45], [46] for a general class of systems in $\mathbf{R}^n$ with analogous Douglis-Nirenberg indices

$$s_1 = \dots = s_{k-1} = 0, \; s_N = -1$$

$$t_1 = \dots = t_{N-1} = 2, \; t_N = 1$$

(N number of components of u). The estimates for general elliptic systems and general boundary conditions are not known.

5. THE BALAYAGE-PRINCIPLE (SWEEPING-OUT PROCESS)

In studying properly posed linear boundary value problems, we notice two facts. First, the a priori estimates, from which the existence and the integral representation follow. Second, the use of potentials. The set of continuous potentials is an adapted set, which describes in a certain sense the regularity of the boundary.

The famous balayage - principle (for the Laplace equation) represents these two facts and is in a certain sense equivalent to the Dirichlet problem. The kernels considered are positive kernels, which are lower semicontinuous.

Definition 1: *A kernel $\Phi \geqq 0$ satisfies the balayage - principle, if for every compact set $K \subset \mathbf{R}^n$ of positive Φ - capacity and every measure $\nu \geqq 0$ a positive measure μ with $\operatorname{supp} \mu \subset K$ exists so that*

(1) $\tilde{\Phi}\nu(z) = \tilde{\Phi}\mu(z)$ *on* K, *except a set of Φ-capacity zero,*

(2) $\tilde{\Phi}\nu(x) \geqq \tilde{\Phi}\mu(x)$ *in the whole space* $\mathbf{R}^n$.

We know the balayage - principle for example for elliptic equations of the second order for which the maximum-minimum-principle holds and for parabolic equations of the second order with the same property. The balayage - principle is equivalent to the following integral form of the balayage - principle (see G. ANGER [6] - [9]).

Theorem 1: *The relations (1) and (2) are equivalent to the two relations*

(3) $\int f d\nu = \int f d\mu$ *for every* $f \in \bar{D}(K; \Phi) \subset C(K)$,

(4) $\int g d\nu \geqq \int g d\mu$ *for every* $g = \Phi\lambda$, $\lambda \in \mathfrak{F}^+(\Phi)$.

We know also a balayage - principle for elliptic equations of order 2p. In this case we take for ν a system $(\nu_\alpha)_{|\alpha| \leqq m-1}$ of measures ν_α and the potentials $\Phi\lambda$, $\lambda = (\lambda^\beta)_{|\beta| \leqq p-1}$, $\lambda^\beta \in \mathfrak{F}(\Phi; p-1)$. The relation (1) assumes the form

(5) $\tilde{\Phi}\nu(z) = \tilde{\Phi}\mu(z)$ on K, except a set of Φ-capacity zero,

and the relation (2) the form

(6) $\nu(\Phi\lambda) = \mu(\Phi\lambda)$ for every $\lambda \in \mathfrak{F}(\Phi; p-1)$,

see formulas (15) - (20) of section 3. For the inequality (2) there exists no similar relation. Also for general boundary conditions an abstract balayage - principle exists (see G. ANGER [8], [9], B.-W. SCHULZE [43], [44], [46], G. WILDENHAIN [49], G. ANGER, B.-W. SCHULZE and G. WILDENHAIN [15]).

6. INVERSE PROBLEMS IN POTENTIAL THEORY

We consider the inverse problem of 1.3 only for the Laplace equation. We can formulate a similar problem for the heat equation. However, no results obtained by our method are known for the heat equation. For applications of the inverse problem see M.M. LAVRENTIEV [31], A.I. PRILEPKO [37], A.N. TYCHONOV, V.K. IVANOV and M.M. LAVRENTIEV [47].

Let $\Omega \subset \mathbf{R}^n$, $n \geqq 3$ be a bounded domain with only regular boundary points (stable in the sense of Keldysh) and $\mu \geqq 0$ a measure with supp $\mu \subset \partial\Omega$. Let $\mathfrak{M}_\mu$ be the set of all positive measures $\nu \geqq 0$ with supp $\nu \subset \bar{\Omega}$ satisfying (1), (2) or (3), (4).

Because of the regularity of $\partial\Omega$ the potentials considered are equal on $\partial\Omega$. In the theory of inverse problems two problems are important.

Problem 1: To describe the structure of the convex set $\mathfrak{M}_\mu$ and to find further conditions on ν characterizing subsets of $\mathfrak{M}_\mu$ which are important in applications.

Problem 2: Let μ_1 and μ_2 be two positive measures. What are the conditions on μ_1 and μ_2, such that the potentials $\Phi\mu_1$ and $\Phi\mu_2$ are different outside supp $\mu_1 \cup$ supp μ_2, i.e. $\Phi\mu_1(x^0) \neq \Phi\mu_2(x^0)$ at least at one point $x^0 \notin$ supp $\mu_1 \cup$ supp μ_2.

There are only weak results for Problem 2 (see G. ANGER [14], A.I. PRILEPKO [37]). Let us make some remarks on this problem. Let K = supp $\mu_1 \cup$ supp μ_2 and let γ be a positive measure on supp $\mu_1 \cap$ supp μ_2. Then we can find a measure ν_1 on ∂(supp μ_1) with

$$\Phi\gamma(z) = \Phi\nu_1(z) \text{ outside K}$$

and also another measure ν_2 on ∂(supp μ_2) with

$$\Phi\gamma(z) = \Phi\nu_2(z) \text{ outside K.}$$

These two different measures produce the same potentials outside K. It seems that we must first know the structure of $\mathfrak{M}_\mu$ and only then we can study Problem 2. If $\Phi\mu_1 = \Phi\mu_2$ outside K, then we have a swept-out measure μ on ∂K with the same potential outside K. With the aid of the structure of $\mathfrak{M}_\mu$ (extreme points) we have to prove whether μ_1 and μ_2 can produce the same potential outside K or not.

We consider the space $C(\bar{\Omega})$ and its dual space $C'(\bar{\Omega})$. The space $C'(\bar{\Omega})$ is furnished with the weak topology. We have $\mathfrak{M}_\mu \subset C'(\bar{\Omega})$ and consider on $\mathfrak{M}_\mu$ the induced (weak) topology. For the results which follow, see G. ANGER [14] and G. ANGER and B.-W. SCHULZE [16].

Theorem 1: *$\mathfrak{M}_\mu$ is a convex, (weakly) compact set with $\|\nu\| = \|\mu\|$ for every $\nu \in \mathfrak{M}_\mu$.*

Later we shall discuss the extreme points of $\mathfrak{M}_\mu$, denoted by $\partial_e \mathfrak{M}_\mu$. The convex hull of $\partial_e \mathfrak{M}_\mu$ is dense in $\mathfrak{M}_\mu$ (theorem of Krein-Milman). There is a stronger result on the set $\mathfrak{M}_\mu$, the so-called Choquet theorem [26]. Let $\mathscr{C}$ be an affine function on $\mathfrak{M}_\mu$, i.e.

$$\mathscr{C}(t\nu_1 + (1 - t)\nu_2) = t\mathscr{C}(\nu_1) + (1 - t)\mathscr{C}(\nu_2)$$

for measures ν_1, ν_2, and let $C(\mathfrak{M}_\mu)$ be the space of continuous functions on $\mathfrak{M}_\mu$ and $C'(\mathfrak{M}_\mu)$ the dual space of $C(\mathfrak{M}_\mu)$. According to G. Choquet every measure $\nu \in \mathfrak{M}_\mu$ has the form

$$(1) \qquad \mathscr{C}(\nu) = \int \mathscr{C}(\nu)\,dm(\nu) ,$$

where $m \in C'(\mathfrak{M}_\mu)$ is a positive measure with $\|m\| = 1, m(\mathfrak{M}_\mu \setminus \partial_e \mathfrak{M}_\mu) = 0$, and $\mathscr{C} \in C(\mathfrak{M}_\mu)$ is an affine function. Affine functions on $C(\mathfrak{M}_\mu)$ are for instance

$$\nu \mapsto \mathscr{C}_f(\nu) = \int f\,d\nu , \quad f \in C(\bar{\Omega}) .$$

Now we study the inverse problem in the space $\mathbf{R}^1$. The fundamental solution has the form

$$(2) \qquad \Phi(x,y) = E(x - y) = (x - y)\,\Theta\,(x - y) + c_1(x - y) + c_2 .$$

With the aid of the conditions $\Phi(0,y) = \Phi(1,y) = 0$ we can determine the two constants c_1 and c_2. The intervals (a,b) are subsets of $[0,1]$. Let μ be a measure on the boundary $\{a,b\}$

$$\mu = s\delta_a + (1 - s)\,\delta_b .$$

Then there exists a point $x^o = b - (b - a)s$, the barycentre, such that

$$\Phi\delta_{x^o}(z) = \Phi\mu(z) \text{ for all } z \notin (a,b) ,$$

$$\Phi\delta_{x^o}(x) \geqq \Phi\mu(x) \text{ for all } x \in \mathbf{R}^1 .$$

Let Ω be a domain in $\mathbf{R}^n$. For an arbitrary measure $\mu \geqq 0$ on $\partial\Omega$ there is generally no point $x^o \in \Omega$ with the above properties. Extreme points of $\mathfrak{M}_\mu$ in $\mathbf{R}^1$ are the measures

$$\delta_{x^o,\alpha,\beta} = \frac{\beta}{\alpha+\beta}\,\delta_{x^o-\alpha} + \frac{\alpha}{\alpha+\beta}\,\delta_{x^o+\beta}\ ,$$

$0 < \alpha \leqq x^o - a$, $0 < \beta \leqq b - x^o$. Every positive measure m on the set

$$A = \{x : 0 < \alpha \leqq x^o - a,\ \ 0 < \beta \leqq b - x^o\}$$

with $m(A) = 1$ produces a measure $\nu \in \mathfrak{M}_\mu$, where

$$(3) \qquad \nu(f) = \int \delta_{x^o,\alpha,\beta}(f)\,dm(\alpha,\beta)\ .$$

Sometimes it is interesting to consider the inverse problem with respect to the kernel Φ with $\Phi(x,y) = \Theta(x - y)$, where $\mu = \delta_{x^o}$ and $\mathfrak{M}_{\delta_{x^o}}$ is the set of all positive measures ν with compact support and

$$\Phi\nu(x) = \Phi\delta_{x^o}(x) \text{ for all } x > x^o\ .$$

In the sequel we take $u = \mu_{x^o}$, $x^o \in \Omega \subset \mathbf{R}^n$. In this case we can characterize classes of extreme points. (For the following theorems see G. ANGER and B.-W. SCHULZE [16].)

Theorem 2: *Let $\Omega' \subset \Omega$ be a simply connected domain containing the point x^o and suppose that the boundary $\partial\Omega'$ consists only of regular (stable) points. Let $\mu^{\Omega'}_{x^o}$ be the harmonic measure with respect to Ω'. Then*

$$\mu^{\Omega'}_{x^o} \in \partial_e\,\mathfrak{M}_{\mu_{x^o}}$$

Theorem 3: *If the domain $\Omega' \subset \Omega$ is multi-connected or $\partial\Omega \neq \partial\bar{\Omega}$ and $\nu \in \mathfrak{M}_{\mu_{x^o}}$ with* supp $\nu = \partial\Omega'$ *, then $\nu \notin \partial_e\,\mathfrak{M}_{\mu_{x^o}}$.*

This theorem follows from the fact that there exists a part $K \subset \partial\Omega'$ not contained in $\partial\Omega_\infty$, where Ω_∞ is the component of $\mathbf{R}^n\backslash\Omega'$ containing the point ∞ . Let the restriction of ν to K be ν_1 and the swept-out measure on $\partial\Omega_\infty$ be ν_1'. Let the restriction of ν to $\partial\Omega_\infty$ be ν_2. Then there exists a positive constant ε, such that $\nu_2 - \varepsilon\nu_1'$ is a positive measure. Then

$$(4) \qquad \gamma_1 = \nu - \varepsilon\nu_1 + \varepsilon\nu_1' \,, \quad \gamma_2 = \nu + \varepsilon\nu_1 - \varepsilon\nu_1'$$

are elements of $\mathfrak{M}_{\mu_{x^0}}$ and $\nu = \frac{1}{2}(\gamma_1 + \gamma_2)$.

Theorem 4: *The set* $\partial_e \mathfrak{M}_{\mu_{x^0}}$ *is not closed.*

To prove this theorem, we can take a simply connected domain $\Omega' \subset \Omega$, which contains points z not in $\partial(\bar{\Omega}')$, for instance points in a hyperplane. For this domain Ω' we can consider the Wiener procedure $\Omega_k \subset \Omega'$ with $\bigcup \Omega_k = \Omega'$. The harmonic measures $\mu_x^{\Omega_k}$ are elements of $\partial_e \mathfrak{M}_{\mu_{x^0}}$, but by Theorem 3 $\mu_x^{\Omega'}$ is not an element of $\partial_e \mathfrak{M}_{\mu_{x^0}}$.

Now we construct other extreme points. For example, let us consider a ball

$$K_r = \{x : |x - x^*| < r\} .$$

Let $x^0 \in K_r$, where x^0 can be the point x^*. We consider the harmonic measure $\mu_{x^0}^{K_r}$. This measure has the form (see formula (16) of 1.2)

$$\mu_{x^0}^{K_r}(f) = \int f(y)\,d\mu_{x^0}^{K_r}(y) = \int\limits_{\partial K_r} f(y)\,\frac{(r^2 - |x^0 - x^*|^2)\,dO(y)}{r\omega_n |x^0 - y|^n}$$

Let z be another point in K_r. Then we have

$$\delta_z(f) = \mu_z^{K_r}(f)$$

for all harmonic functions $f \in C(\bar{\Omega})$. We consider a sufficiently small constant a such that

$$g_a(y) = \frac{r^2 - |x^0 - x^*|^2}{r\omega_n |x^0 - y|^2} - a\,\frac{r^2 - |z - x^*|^2}{r\omega_n |z - y|^n} \geqq 0 \quad \text{on} \quad \partial K_r$$

In the case $x^0 = x^*$ we get

$$g_a(y) = \frac{1}{r\omega_n}\left(\frac{1}{r^{n-2}} - a\,\frac{|z - x^*|^2}{|z - y|^n}\right) .$$

We obtain in this notation

$$\mu_{x^0}^{K_r}(f) - a\,\mu_z^{K_r}(f) = \int_{\partial K_r} f(y)g_a(y)dO(y) .$$

Every

$$\nu \in \mathfrak{M}_{\mu_{x^0}^{K_r}} \quad \text{with supp } \nu \subset \partial K_r \cup \{z\}$$

must be of the form $a\delta_z + \gamma$, where supp $\gamma \subset \partial K_r$, $\gamma \geqq 0$ and $a \geqq 0$. Let $f \in C(\bar{\Omega})$ be a harmonic function. From

$$a\delta_z(f) + \gamma(f) = a\mu_z^{K_r}(f) + \gamma(f) = \mu_{x^0}^{K_r}(f)$$

it follows that

$$\gamma(f) = \mu_{x^0}^{K_r}(f) - a\mu_z^{K_r}(f) = \int_{\partial K_r} f(y)g_a(y)dO(y) .$$

The measure γ is uniquely determined, because the Dirichlet problem is solvable for every $f \in C(\partial\Omega)$.

There exists greatest constant a_0 such that $g_{a_0} \geqq 0$ on ∂K_r and $g_{a_0}(y^0) = 0$ in at least one point $y^0 \in \partial K_r$. The measure

$$(5) \qquad \nu_{a_0} = \mu_{x^0}^{K_r} + a_0\delta_z - a_0\mu_z^{K_r} = a_0\delta_z + (\mu_{x^0}^{K_r} - a_0\mu_z^{K_r})$$

is an element of $\partial_e \mathfrak{M}_{\mu_{x^0}^{K_r}}$. In the following we shall give a short proof of this fact. For $a > a_0$ we have $g_a^- \neq 0$. This means, that ν_a is not positive and therefore not an element of $\mathfrak{M}_{\mu_{x^0}^{K_r}}$. If ν_{a_0} is no extreme point, then

$$(6) \qquad \nu_{a_0} = t\gamma_1 + (1 - t)\gamma_2, \quad 0 < t < 1 ,$$

where

$$\gamma_1(f) = \int_{\partial K_r} g_{a_1}(y)f(y)dO(y), \quad a_1 < a_0 ,$$

$$\gamma_2(f) = \int_{\partial K_r} g_{a_2}(y)f(y)dO(y), \quad a_2 < a_0 .$$

We can construct a sequence of functions f_k such that

$$g_{a_0}(y^0) = t\, g_{a_1}(y^0) + (1-t)\, g_{a_2}(y^0) .$$

Since $o < t < 1$, we get

$$g_{a_0}(y^0) = g_{a_1}(y^0) = g_{a_2}(y^0) = 0 .$$

For

$a_1 < a_0$ and $a_2 < a_0$ we have

$g_{a_1}(y^0) > 0$ and $g_{a_2}(y^0) > 0$.

Therefore the relation (6) is impossible. The measure ν_{a_0} is an extreme point.

Theorem 5: *The measures* ν_{a_0} *are elements of* $\partial_e \mathfrak{M}_{\mu_{x^0}}^{K_r}$.

Let ν be a positive measure with finite energy. Then the following relation holds

$$I(\nu) = \int\left(\int \Phi(x,y)\,d\nu(y)\right)d\nu(x) = c_n \int_{\mathbf{R}^n} (\operatorname{grad}(\Phi\nu))^2\, dx .$$

The convexity of I yields for the subset M of measures $\nu \in \mathfrak{M}_{\mu_{x^0}}$ with finite energy and supp $\nu \subset K$

Theorem 6: *Let* $\sup_{\nu \in M} I(\nu) = I(\nu^0)$, *where* $\nu^0 \in M$. *Then* ν^0 *is an extreme point of* $\mathfrak{M}_{\mu_{x^0}}$.

In the general case we can prove with the aid of the special measures ν_{a_0}

Theorem 7: *It holds* $\sup\{I(\nu);\ \nu \in \mathfrak{M}_{\mu_{x^0}}\} = \infty$

Using the harmonic polynomials, we get

Theorem 8: *The measures* $\nu \in \mathfrak{M}_{\mu_{x^0}}$ *have the same barycentre and the same moments of higher order.*

The following theorem was proved by A. CORNEA

Theorem 9: *Let Φ be the Newton kernel and F the set of all linear combinations of functions* $\Phi(x,z^{(k)})$, $z^{(k)} \in \mathbf{R}^n \setminus \bar{\Omega}$. *A measure* $\nu \in \mathfrak{M}_{\mu_{x^0}}$ *is an element of* $\partial_e \mathfrak{M}_{\mu_{x^0}}$ *if and only if* F *is dense in* $L^1(\nu)$.

REFERENCES

[1] S. Agmon, Maximum properties for solutions of higher order elliptic equations. Bull. Amer. math. Soc. 66, 77-80 (1960).

[2] S. Agmon, A. Douglis and L. Nirenberg, Estimates near the boundary for solutions of elliptic partial differential equations satisfying general boundary conditions, I. II. Comm. pure appl. Math. 12, 623-727 (1959), 17, 35-92 (1964).

[3] G. Albinus, Bemerkungen zu einer potentialtheoretischen Methode. Math. Nachr. 55, 381-397 (1972).

[4] G. Albinus, Über das im Sinne von N. Wiener verallgemeinerte Dirichletsche Problem für die Helmholtzsche Schwingungsgleichung in beliebigen beschränkten Gebieten (to appear in Math. Nachr.).

[5] G. Albinus, N. Boboc und P. Mustaţa, A priori evaluations for the Dirichlet problem associated with a linear elliptic operator (to appear in Math. Nachr.).

[6] G. Anger, Stetige Potentiale und deren Verwendung für einen Neuaufbau der Potentialtheorie. Dissertation, Technische Hochschule Dresden, 1957.

[7] G. Anger, Sur le rôle des potentiels continues dans les fondements de la théorie du potentiel. Séminaire de théorie du potentiel, 2ème année, 1957/58, Institut H. Poincaré, Université de Paris.

[8] G. Anger, Balayage-Prinzip und stetige Projektionen. Wiss. Z. Techn. Universität Dresden 11, 417-426 (1962).

[9] G. Anger, Funktionalanalytische Betrachtungen bei Differentialgleichungen unter Verwendung von Methoden der Potentialtheorie, I. Akademie-Verlag, Berlin 1967.

[10] G. Anger, Eine allgemeine Kerntheorie, I. S^p-Kerne. Math. Nachr. 37, 153--176 (1968).

[11] G. Anger, Potentialtheoretische Untersuchungen über Randwertaufgaben, deren Randoperatoren Ableitungen beliebiger Ordnung enthalten. Kolloquiumsbericht "Elliptische Differentialgleichungen, II.", Berlin 1971, S. 15-78.

[12] G. Anger, Eindeutigkeitssätze und Approximationssätze für Potentiale, I. Math. Nachr. 43, 123-142 (1970).

[13] G. Anger, Eindeutigkeitssätze und Approximationssätze für Potentiale, II. Math. Nachr. 50, 229-244 (1971).

[14] Г. Ангер, Теоремы единственности и аппроксимации для потенциалов, III. обратные задачи (to appear in Сибирский Математический Журнал).

[15] G. Anger, B.-W. Schulze and G. Wildenhain, Neuere Entwicklungen auf dem Gebiet der Potentialtheorie (to appear in "Die Entwicklung der Mathematik in der DDR", Berlin 1974).

[16] G. Anger and B.-W. Schulze, Some remarks on harmonic measures and inverse problems (to appear in "Beiträge zur Analysis und Angewandten Mathematik", Berlin 1974).

[17] М.А. Атаходжаев, О корректности по Тихонову постановки одной задачи для уравнения $\Delta^2 \mu = 0$. Краев. задачи для дифференц. уравнений с частными производными, фан, 1970, 81-83.

[18] H. Bauer, Harmonische Räume und ihre Potentialtheorie. Berlin-Heidelberg-New York 1966.

[19] N. Boboc et P. Mustaţâ, Espaces harmoniques associés aux opérateurs différentiels linéaires du second ordre de type elliptique. Berlin-Heidelberg-New York 1968.

[20] M. Brelot, Eléments de la théorie classique du potentiel. Centre de Documentation Univ., 4e édit., Paris 1969.

[21] M. Brelog (editor), Potential theory. Centro Internazionale Matematico Estivo. Edizioni Cremonese, Roma 1970.

[22] M. Brelot, Les étapes et les aspects multiples de la théorie du potentiel. Enseignement math., II. Sér. 18, 1, 1-36 (1972).

[23] F.E. Browder, Functional analysis and partial differential equations, II. Math. Ann. 145, 81-226 (1962).

[24] A. Canfora, Teorema del massimo modulo e teorema di esistenza per il problema di Dirichlet relativo di sistemi fortemente ellitici. Ricerche Mat. 15, 249-294 (1966).

[25] G. Choquet, Theory of capacities. Ann. Inst. Fourier 5, 131-295 (1953/54).

[26] G. Choquet, Lectures on analysis II, New York-Amsterdam, 1969.

[27] C. Constantinescu and A. Cornea, Potential theory and harmonic spaces. Berlin-Heidelberg-New York 1972.

[28] В.Т. Иванов, Г.П. Смирнов, О прямой и обратной краевых задачах для уравнения теплопроводности. Дифференц. уравнения, 8, 2023-2028 (1972).

[29] Ю.П. Красовский, Исследование потенциалов связанных с краевыми задачами для эллиптических уравнений. Изд. АН СССР, сер. матем. 31, 587-640 (1967).

[30] Е.М. Ландис, Уравнения второго порядка эллиптического и параболического типов. Москва 1971.

[31] M.M. Lavrentiev, Some improperly posed problems of mathematical physics. Berlin-Heidelberg-New York 1967.

[32] M.M. Lavrentiev, V.O. Romanov and V.G. Vasiliev, Multidimensional inverse problems for differential equations. Berlin-Heidelberg-New York 1970.

[33] J. Lukeš, A new type of generalized solution of the Dirichlet problem for the heat equation. Nonlinear Evolution Equations and Potential Theory, Academia Praha 1974.

[34] В.Г. Мазья, В.П. Хавин, Нелинейная теория потенциала. Успехи мат. наук 27, 67-138 (1972).

[35] S.G. Michlin, Lehrgang der mathematischen Physik. Berlin 1972 (Übers. a.d. Russischen).

[36] C. Miranda, Teorema del massimo modulo e teorema di esistenza e di unicetà per il problema Dirichlet relativo alle equazioni ellitiche in due variabili. Ann. Mat. pura appl., IV. Ser., 46, 265-311 (1958).

[37] A.I. Prilepko, Über Existenz und Eindeutigkeit von Lösungen inverser Probleme der Potentialtheorie (to appear in Math. Nachr.).

[38] В.Г. Романов, Некоторые обратные задачи для уравнений гиперболического типа. Академия наук СССР, Сибирское отделение, Новосибирск 1972.

[39] M. Schechter, On L^p estimates and regularity I, II, III. Amer. J. Math. 85, 1-13 (1963), Math. Scandinav. 13, 47-69 (1963), Ricerche Mat. 13, 192-206 (1964).

[40] B.-W. Schulze, Potentiale bei der Wellengleichung im R^2 und Charakterisierung der Mengen der Kapazität Null. Kolloquium "Elliptische Differentialgleichungen I", Berlin 1970, S. 137-157.

[41] B.-W. Schulze, Mengen der Kapazität Null für nichtelliptische Differentialgleichungen. Das Dirichlet-Problem für $u_{xy} = 0$. Kolloquium "Elliptische Differentialgleichungen II", Berlin 1971, S. 217-246.

[42] B.-W. Schulze, Nullmengensysteme in der Potentialtheorie. Math. Nachr. 49, 293-309 (1971).

[43] B.-W. Schulze, Ein Balayage-Prinzip für korrekte Randwert-Probleme I. Math. Nachr. 56, 307-344 (1973).

[44] B.-W. Schulze, Ein Balayage-Prinzip für korrekte Randwert-Probleme II. (to appear in Math. Nachr.).

[45] Б.-В. Шулце, Об априорных оценках в равномерных нормах для сильно эллиптических систем (to appear in Сибирский Математический Журнал).

[46] B.-W. Schulze, Anwendungen des Balayage-Prinzipes auf allgemeine Randwert--Probleme für Systeme partieller Differentialgleichungen, Thesis, Universität Rostock, 1973.

[47] А.Н. Тихонов, В.К. Иванов, М.М. Лаврентьев, Некорректно поставленные задачи. Дифференциальные уравнения с частными производными, 224-238, Москва 1970.

[48] G. Wildenhain, Eine Charakterisierung von Mengen der Kapazität Null im R^1. Math. Nachr. 32, 269-275 (1966).

[49] G. Wildenhain, Potentialtheorie linearer elliptischer Differentialgleichungen beliebiger Ordnung. Berlin 1968.

[50] G. Wildenhain, Vergleich verschiedener Kapazitätsbegriffe. Kolloquium "Eliptische Differentialgleichungen I", Berlin 1970, S. 179-190.

[51] G. Wildenhain, Das "feine" Dirichlet-Problem für elliptische Differentialgleichungen beliebiger Ordnung. Kolloquium "Elliptische Differentialgleichungen II", Berlin 1971, S. 247-264.

[52] G. Wildenhain, Über das Randverhalten der Hilbertraumlösung des Dirichlet--Problems. Math. Nachr. 48, 227-235 (1971).

[53] Symposium on Non-Well-Posed Problems and Logarithmic Convexity. Lecture Notes in Mathematics, vol. 316, Berlin-Heidelberg-New York 1973.

MLU Halle, Sektion Mathematik, GDR.

REGULARITY RESULTS FOR SOME DIFFERENTIAL EOUATIONS ASSOCIATED WITH MAXIMAL MONOTONE OPERATORS IN HILBERT SPACES

Viorel Barbu
Iaşi (Romania)

1. INTRODUCTION

Let H be a real Hilbert space whose norm and inner product is denoted respectively by $|\ |$ and $(\ ,\ )$. A subset $A \subset H \times H$ is called monotone if

$$(y_1 - y_2,\ x_1 - x_2) \geq 0 \quad \text{for all } [x_i,\ y_i] \in A, \quad i = 1,2\ .$$

A monotone set is maximal monotone if it is not properly contained in any monotone subset of H x H. Equivalently, $R(I + \lambda A) = H$ for all $\lambda > 0$.[1)]

We consider the Cauchy problem

$$(1.1) \qquad \begin{aligned} &\frac{du}{dt}(t) + Au(t) \ni f(t), \quad \text{a.e. } 0 < t < T \\ &u(0) = x\ , \end{aligned}$$

[1)] If A,B are subsets of H x H then the following notations will be used:

$D(A) = \{x \in H;\ [x,y] \in A \ \text{ for some } y \in H\}$,
$R(A) = \{y \in H;\ [x,y]\ \ A \ \text{ for some } x \in H\}$,
$\lambda A = \{[x, \lambda y] \in H \times H;\ [x,y] \in A\}$, λ a real number,
$A+B = \{[x,y+z] \in H \times H;\ [x,y] \in A,\ [x,z] \in B\}$,
$A^{-1} = \{[y,x] \in H \times H;\ [x,y] \in A\}$.
Denote $I = \{[x,x] \in H \times H;\ x \in H\}$.

Any $A \subset H \times H$ may be considered as a multivalued mapping from H into itself defined by

$Ax = \{y;\ [x,y] \in A\}$ for $x \in D(A)$

(reviewers' remark).

where $[0,T]$ is a fixed real interval. For every $1 \leq p \leq \infty$ we denote by $W^{1,p}(0,T;H)$ the space of all absolutely continuous functions $u : [0,T] \mapsto H$ such that $du/dt \in L^p(0,T;H)$.

It is well known (cf [6], [9], [10]) that for every $x \in D(A)$ and $f \in W^{1,1}(0,T;H)$ the equation (1.1) has a unique solution $u \in W^{1,\infty}(0,T;H)$. In particular, it follows that there exists a semigroup of nonlinear contractions $S(t)$ defined on $\overline{D(A)}$ such that

$$\frac{d}{dt} S(t)x + A^o S(t)x = 0, \quad \text{a.e. } t > 0$$

for every $x \in D(A)$. Here A^o denotes the minimal section of A, i.e. $|A^o x| = \inf\{|y| ; y \in Ax\}$. If $x \in \overline{D(A)}$ and $f \in L^1(0,T;H)$ the equation (1.1) has a unique weak solution $u(t)$ which is continuous on $[0,T]$ (see [6]).

More precisely, there exists a unique function $u \in C(0,T;H)$ such that $u(0) = x$ and

$$\frac{1}{2}|u(t) - x|^2 \leq \frac{1}{2}|u(s) - u|^2 + \int_s^t (f(\tau) - v, \; u(\tau) - u)d\tau$$

for every $[u,v] \in A$ and all $0 < s < t < T$.

Here we present some relevant classes of nonlinear maximal monotone operators A with the property that the weak solutions of (1.1) (with f in a suitable class of integrable functions) are absolutely continuous on $]0,T[$ and satisfy (1.1) a.e. on $]0,T[$. Moreover, it turns out that the semigroups $S(t)$ generated by these operators have a smoothing effect on initial data, i.e., $S(t)\,\overline{D(A)} \subset D(A)$ for all $t > 0$ and

$$t \frac{d}{dt} S(t)x \in L^\infty(0,T;H)$$

for every $x \in \overline{D(A)}$. It is well known that in linear case this property is characteristic for analytic semigroups and their generators A may be described in a simple manner. In nonlinear case only sufficient conditions are known. The characterization of nonlinear maximal monotone operators in Hilbert spaces with this property appears to be a formidable task.

2. DIFFERENTIAL EQUATIONS ASSOCIATED WITH SUBGRADIENT MAPPINGS

Let φ be a convex and lower semicontinuous function from H into $]-\infty,\infty[$, $\varphi \not\equiv +\infty$.
We set

$$D(\varphi) = \{x \in H;\quad \varphi(x) < +\infty\}$$

and denote by $\partial\varphi(x)$ the set of all $y \in H$ such that

$$\varphi(x) \leq \varphi(u) + (x-u,y) \quad \text{for all } u \in H .$$

The set $\partial\varphi \subset H \times H$ is called the subdifferential of φ and every $y \in \partial\varphi(x)$ is said to be a subgradient of φ at x. It is well known (see e.g. [6], [16]) that $\partial\varphi$ is maximal monotone and $D(\partial\varphi)$ is dense in $D(\varphi)$.

Let us mention that a linear operator is the subdifferential of a lower semicontinuous convex function iff it is selfadjoint.

The following result was given by H. BRÉZIS [4].

Theorem 1: *Let* φ *be a lower semicontinuous convex function from* H *into* $]-\infty, +\infty]$, *nonidentically* $+\infty$. *Let* $A = \partial\varphi$. *Then for every* $f \in L^2(0,T;H)$ *and* $x \in \overline{D(A)}$ *there exists a unique function* $u \in C(0,T;H)$ *satisfying*

(2.1) $u(t)$ *is absolutely continuous on* $]0,T[$

(2.2) $u(t) \in D(A)$ a.e. $t \in]0,T[$

(2.3) $\sqrt{t}\, du/dt \in L^2(0,T;H)$

(2.4) $\frac{du}{dt}(t) + (Au(t) - f(t))^0 = 0$ a.e. $t \in]0,T[$

(2.5) $u(0) = x.$

Moreover, if $x \in D(\varphi)$ *then* $du/dt \in L^2(0,T;H)$.

For the proof of this remarkable result we refer to [4], [6]. In particular, Theorem 1 implies

Corollary 1: *Let* $A = \partial\varphi$ *and let* $S(t)$ *be the semigroup generated by* A *on* $\overline{D(A)}$.

Then $S(t)\,\overline{D(A)} \subset D(A)$ *for every* $t > 0$ *and*

(2.6) $$|A^o S(t)x| \leq |A^o u_o| + \frac{1}{t}|x-u_o|$$

for every $x \in \overline{D(A)}$, *every* $u_o \in D(A)$ *and all* $t > 0$.

It is interesting that the above results remain true for "small" perturbations of A by maximal monotone operators, i.e., for A + B where $D(A) \subset D(B)$ and $|B^o u| \leq \alpha |A^o u| + C$ locally in D(A) where $0 < \alpha < 1/2$ (see [4]).

These results have many applications in the theory of existence and regularity of solutions for unilateral problems of parabolic and hyperbolic type (see [4], [7]).

3. THE SQUARE ROOT OF A NONLINEAR MAXIMAL MONOTONE OPERATOR

We begin with the following existence result we have proved in [1] (see also [2]).

Let A be a maximal monotone subset of H x H such that $0 \in R(A)$. Then for every $x \in D(A)$ the boundary problem

(3.1) $$\frac{d^2u}{dt^2} \in Au \quad \text{a.e.} \quad t > 0$$

(3.2) $$u(0) = x, \quad \sup_{t \geq 0} |u(t)| < \infty$$

has a unique solution $u \in W^{2,2}_{loc}(0, \infty; H)$.
Moreover, $\frac{d^2u}{dt^2} \in L^2(0, \infty; H)$ and $\sqrt{t}\,\frac{du}{dt} \in L^2(0, \infty; H)$.

For every $x \in D(A)$ denote by $u_x(t)$ the corresponding solution of the problem (3.1), (3.2). We see that $|u_x - u_y| \leq |x-y|$ for all $t \geq 0$ and $x, y \in D(A)$ and denote by $S_{1/2}(t)$ the extension of the mapping $x \mapsto u_x(t)$ from $\overline{D(A)}$ into itself. It is easy to see that $S_{1/2}(t)$ is a semigroup of nonlinear contractions on $\overline{D(A)}$. Thus according to a well known result of Y. KOMURA [12] (see also M. CRANDALL and A. PAZY [9]) there exists a unique maximal monotone set $A_{1/2} \subset H \times H$ such that $D(A_{1/2}) \subset \overline{D(A)}$ and $\frac{d}{dt}S_{1/2}(t)x+A^{o}_{1/2}S(t)x= 0$ a.e. on $]0,\infty[$ for all $x \in D(A_{1/2})$.

If A is linear then according to a well known result of Balakrisnan (see e.g. [13]) $A_{1/2}$ coincides with the square root of A. We call the operator $A_{1/2}$ the square root of A. This denomination is entitled by some remarkable properties which recall the square root of a liner maximal monotone operator (see H. BRÉZIS [5], Y. KONISHI [11]). We only notice the convexity inequality (see [5])

$$|A^{o}_{1/2}\, x| \le 2|A^{o}x|^{1/2}|x|^{1/2}\ , \quad x \in D(A)\ .$$

If $A = \partial\varphi$ where $\varphi : H \mapsto]0, +\infty]$ is lower semicontinuous and convex such that $\varphi(0) = 0$ then (see [1], [5]) $D(A_{1/2}) = D(\varphi)$ and

$$(3.3) \quad \tfrac{1}{2}|A^{o}_{1/2}u|^{2} = \varphi(u) \quad \text{for all } u \in D(\varphi)\ .$$

It appears that the semigroup $S_{1/2}(t)$ has remarkable regularity properties. More precisely, the following result was proved in [1] and [5].

Theorem 2: *Let* A *be a maximal monotone subset of* $H \times H$ *such that* $0 \in R(A)$ *and let* $A_{1/2}$ *be its square root. Then*
(i) $S_{1/2}(t)\,\overline{D(A)} \subset D(A_{1/2})$ *for all* $t > 0$. *In addition, for every* $x \in \overline{D(A)}$, $t^{1/2}\,\frac{d}{dt}\,S_{1/2}(t)x \in L^{2}(0,\infty\,;H)$, $t^{3/2}\,\frac{d^{2}}{dt^{2}}\,S_{1/2}(t)x \in L^{2}(0,\infty\,;H)$, *and*

$$(3.4) \quad \left|\frac{d}{dt}\,S_{1/2}(t)x\right| \le |x-y|/t \quad \textit{for all } t > 0 \textit{ and } y \in A^{-1}(0)\ .$$

(ii) *For every* $x \in D(A_{1/2})$ *the function* $t \mapsto S_{1/2}(t)x$ *is continuously differentiable on* $[0,\infty[$ *and*

$$(3.5) \quad t^{1/2}\,\frac{d^{2}}{dt^{2}}\,S_{1/2}(t)x \in L^{2}(0,\infty\,;H)\ .$$

(iii) *If* $x \in D(A)$ *then* $\frac{d^{2}}{dt^{2}}\,S_{1/2}(t)x \in L^{2}(0,\infty\,;H)$.

To apply Theorem 2 it is necessary to recognize the operators which are square roots of maximal monotone subsets of H x H. For cyclically maximal monotone operators (i.e., for subdifferentials of lower semicontinuous convex functions on H) the answer is given by the following theorem.

Theorem 3: *Let* A *be a cyclically maximal monotone subset of* H x H *such that* $0 \in R(A)$. *If the function* $u \mapsto \frac{1}{2}|A^o u|^2$ *is convex, then there exists a maximal monotone subset* $B \subset H \times H$ *such that* $A = B_{1/2}$.

P r o o f. It suffices to assume that $0 \in A^o$. Since A is maximal monotone, the convex function $\varphi(u) = \frac{1}{2}|A^o u|^2$ is lower semicontinuous from H into $[0, +\infty]$. Then $B = \partial\varphi$ is maximal monotone and by (3.3) we have $D(B_{1/2}) = D(\varphi)$ and

$$(3.6) \qquad |B^o_{1/2}u| = |A^o u| \quad \text{for all } u \in D(A) .$$

Since $B_{1/2}$ and A are cyclically maximal monotone, (3.6) implies that $A = B_{1/2}$. Indeed, let $\varphi_{1/2}$ and $\psi_{1/2}$ be lower semicontinuous convex functions from H into $[0, +\infty]$ such that $B_{1/2} = \partial\varphi_{1/2}$ and $A = \partial\psi_{1/2}$. We may assume that $\varphi_{1/2}(0) = \psi_{1/2}(0) = 0$. For every $x \in D(\varphi_{1/2})$ there exists a function $u \in C(0,\infty;H)$ such that $\frac{du}{dt} \in L^2(0,\infty;H)$ and (see Theorem 1),

$$\frac{du}{dt} + (\partial\varphi_{1/2})^o u(t) = 0 \quad \text{a.e. } t > 0$$

$$u(0) = x .$$

We have (see [6])

$$\left|\frac{du}{dt}\right|^2 + \frac{d}{dt}\varphi_{1/2}(u) = 0 \quad \text{a.e. } t > 0 .$$

Hence

$$\varphi_{1/2}(x) = \int_0^\infty \left|\frac{du}{dt}\right|^2 dt = \int_0^\infty |A^o u(t)|^2 dt$$

because $\lim_{t\to\infty} \varphi_{1/2}(u(t)) = 0$. On the other hand,

$$\frac{d}{dt}\psi_{1/2}(u(t)) = (A^o u(t), u'(t)) \quad \text{a.e. } t > 0 .$$

Consequently, $x \in D(\psi_{1/2})$ and

$$\psi_{1/2}(x) = \int_0^\infty (A^0 u(t),\ B^o_{1/2}u(t))dt\ .$$

Thus we have proved that $D(\varphi_{1/2}) \subset D(\psi_{1/2})$ and $\varphi_{1/2}(x) \geq \psi_{1/2}(x)$ for every $x \in D(\varphi_{1/2})$. Replacing $\varphi_{1/2}$ by $\psi_{1/2}$, one finally deduces that $\varphi_{1/2} = \psi_{1/2}$. This completes the proof.

4. DIFFERENTIAL EQUATIONS ASSOCIATED WITH MONOTONE HEMICONTINUOUS OPERATORS

Let H be a real Hilbert space and let V be a real reflexive Banach space. Denote by V′ the dual of V and assume that

$$V \subset H \subset V'$$

with each inclusion mapping continuous and densely defined. Let (v', v) be the pairing between v' in V′ and v in V; if $v, v' \in H$, this is the ordinary inner product in H. Denote by $|\ \ |$ the norm in H and by $\|\ \ \|$ and $\|\ \ \|_*$ the norm in V and V′ respectively.

Let A be a nonlinear operator from V into V′ which is assumed to satisfy

(i) A is hemicontinuous (i.e., the function $\lambda \mapsto (A(u+\lambda v),\ w)$ is continuous on $]-\infty, +\infty[$ for all u, v, w in V).

(ii) There exists $\omega > 0$ such that

$$(Au - Av,\ u-v) \geq \omega \|u-v\|^2 \quad \text{for all } u,\ v \in V\ .$$

(iii) There exists $C > 0$ such that

$$\|Au\|_* \leq C(\|u\| + 1) \quad \text{for all } u \in V\ .$$

Let us denote $D(A) = \{u \in V;\ Au \in H\}$. It is easy to see that $\overline{D(A)} =$

= H. Moreover, the restriction of A to D(A) is maximal monotone in H x H (see e.g. [7]).

We consider the problem

$$(4.1)\qquad \frac{du}{dt} + Au = f \quad \text{a.e. } t \in]0,T[$$

$$u(0) = x .$$

It is well known (see J.L. LIONS [14]) that if A is monotone from V into V′, coercive (i.e. $(Au, u) \geq \omega \|u\|^2$, $u \in V$) and satisfies (i) and (iii), then for every $f \in L^2(0,T;V')$ and $x \in H$ the problem (4.1) has a unique solution $u \in C(0,T;H) \cap L^2(0,T;V)$ such that $\frac{du}{dt} \in L^2(0,T;V')$. Moreover, if $x \in D(A)$ and $f \in W^{1,1}(0,T;H)$ it follows from the general existence result presented before that $\frac{du}{dt} \in L^\infty(0,T;H)$ and $Au \in L^\infty(0,T;H)$.

These results are completed by the following theorem (see [3]).

Theorem 4. *Suppose that* A *satisfies assumptions* (i), (ii) *and* (iii). *Let* $f \in L^2(0,T;H)$ *be such that*

$$(4.2)\qquad tf \in L^\infty(0,T;H),\quad t\frac{df}{dt} \in L^2(0,T;V') .$$

Then for every $x \in H$ *the problem* (4.1) *has a unique solution* $u \in C(0,T;H) \cap L^2(0,T;V)$ *satisfying*

$$(4.3)\qquad \frac{du}{dt} \in L^2(0,T;V')$$

$$(4.4)\qquad t\frac{du}{dt} \in L^\infty(0,T;H) \cap L^2(0,T;V) .$$

Remark. In particular, the nonlinear semigroup S(t) generated by A on $\overline{D(A)} = H$ maps H into D(A) for every $t > 0$ and $t \frac{d}{dt} S(t)x \in L^\infty(0,T;H)$.

P r o o f o f T h e o r e m 4. Here we give a sketch of proof, and we refer to [3] for details.

We consider the approximating equations

$$(4.5) \qquad \frac{du}{dt} + A_\lambda u = f, \ 0 \le t \le T$$
$$u(0) = x$$

where $A_\lambda = \lambda^{-1}(I - (I + \lambda A)^{-1})$ is Yosida approximation of $A : D(A) \mapsto H$. Let $u_\lambda \in C^1(0,T;H)$ be the unique solution of (4.5).

Multiplying the equation (4.5) by $\frac{d}{dt}(tA_\lambda u_\lambda)$ one obtains

$$\frac{1}{2}\frac{d}{dt}|tA_\lambda u_\lambda|^2 + t^2\left(\frac{d}{dt} A_\lambda u_\lambda, \frac{du_\lambda}{dt}\right) + t\left(\frac{du_\lambda}{dt}, A_\lambda u_\lambda\right) =$$
$$= \left(tf, \frac{d}{dt}(tA_\lambda u_\lambda)\right) \quad \text{a.e. } t \in]0,T[.$$

We set $V_\lambda(t) = (I+\lambda A)^{-1}u_\lambda(t)$. Using assumption (ii) one gets

$$\omega \left\|\frac{dv_\lambda}{dt}\right\|^2 + \lambda\left|\frac{d}{dt} A_\lambda u_\lambda\right|^2 = \left(\frac{d}{dt} A_\lambda u_\lambda, \frac{du_\lambda}{dt}\right), \quad \text{a.e. } t \in]0,T[.$$

By integrating over $]0,t[$ it follows that

$$|t A_\lambda u_\lambda|^2 + 2\omega\int_0^t \left\|s \frac{dv_\lambda}{ds}\right\| ds + 2\int_0^t s\left(\frac{du_\lambda}{ds}, A_\lambda u_\lambda\right) ds +$$
$$+ 2\lambda\int_0^t \left|s \frac{d}{ds} A_\lambda u_\lambda\right|^2 ds \le 2\int_0^t \left(sf, \frac{d}{ds}(sA_\lambda u_\lambda)\right) ds$$

for all $\lambda > 0$ and $t \in [0,T]$. After some calculations involving this inequality and assumptions (i) and (ii), one finally obtains

$$(4.6) \qquad \begin{aligned} &|t A_\lambda u_\lambda|^2 + \int_0^t \left(\|v_\lambda\|^2 + \left\|s \frac{d}{ds} v_\lambda\right\|^2\right) ds \le C\Big(|x|^2 + \\ &+ |tf|^2 + \int_0^t \left(|f|^2 + \left\|\frac{d}{ds} sf\right\|_*^2\right) ds\Big), \quad 0 \le t \le T. \end{aligned}$$

We first assume that $x \in D(A)$ and $f \in W^{1,2}(0,T;H)$. Then (see e.g. [7], [8]) $\lim_{\lambda\to 0} u_\lambda(t) = \lim_{\lambda\to 0} v_\lambda(t) = u(t)$ uniformly on $[0,T]$ in H, and $\frac{du_\lambda}{dt} \to \frac{du}{dt}$ weakly in $L^2(0,T;H)$. Thus we can pass to the limit in (4.6) to obtain the estimate

$$|t\,Au(t)|^2 + \int_0^t (\|u\|^2 + \|s\,\frac{du}{ds}\|^2)\,ds \le C\,(|x|^2 + |tf|^2 + \tag{4.7}$$
$$+ \int_0^t (|f|^2 + \|\frac{d}{ds}\,sf\|_*^2)\,ds)\,,\quad 0 \le t \le T\,.$$

Let x be arbitrary in H and let $f \in L^2(0,T;H)$ be such that

$t\,\frac{df}{dt} \in D^2(0,T;V')$.

Then there are sequences $\{x_n\} \subset D(A)$ and $\{f_n\} \subset W^{1,2}(0,T;H)$ such that $x_n \to x$ in H, $f_n \to f$ in $L^2(0,T;H)$ and $t\,\frac{df_n}{dt} \to t\,\frac{df}{dt}$ in $L^2(0,T;V')$. Denote by u_n the corresponding solutions of (4.5). It follows then by a standard argument that $u_n(t)$ converges uniformly in H to a solution u of (4.5). The estimate (4.7) clearly implies (4.3) and (4.4).

As an example we consider the boundary problem

$$\frac{du}{dt} - \sum_{i=1}^{n} \partial/\partial x_i a_i(u_x) = f \quad \text{in } \Omega \times]0,T[$$
$$u(x,t) = 0 \qquad \text{on } \Gamma \times]0,T[\tag{4.8}$$
$$u(x,0) = u_o(x) \text{ on } \Gamma$$

where Ω is a bounded domain of $\mathbf{R}^n$ with boundary Γ. Here $u_x = (\frac{\partial u}{\partial x_1}, \ldots \frac{\partial u}{\partial x_n})$ and $a(\xi) = (a_1(\xi), \ldots a_n(\xi))$ define a map from $\mathbf{R}^n$ into itself which is continuous and strictly monotone, i.e.

$$\sum_{i=1}^{n} (a_i(\xi) - a_i(\eta))\,(\xi_i - \eta_i) \ge \omega|\xi - \eta|^2\,, \quad \xi, \eta \in \mathbf{R}^n\,.$$

If

$$|a(\xi)| \le C_1\,|\xi| + C_2 \quad \text{for all } \xi \in \mathbf{R}^n\,,$$

then the operator $A = -\sum_{i=1}^{n} \partial/\partial x_i a_i$ defined from $V = H_0^1(\Omega)$ into $V = H^{-1}(\Omega)$, satisfies the conditions of Theorem 4.

Consequently, for every $u_o \in L^2(\Omega)$ and every $f \in L^2(0,T;L^2(\Omega))$ satisfying

$$tf \in L^{\infty}(0,T;L^2(\Omega)), \quad t\frac{\partial f}{\partial t} \in L^2(0,T;H^{-1}(\Omega))$$

the problem (4.8) has a unique solution $u \in C(0,T;L^2(\Omega)) \cap L^2(0,T;H_0^1(\Omega))$ such that $t\frac{\partial u}{\partial t} \in L^{\infty}(0,T;L^2(\Omega)) \cap L^2(0,T;H_0^1(\Omega))$.

5. THE CASE OF MAXIMAL MONOTONE OPERATORS DOMINATED BY CONVEX FUNCTIONS

Let A be maximal monotone in H x H and let $\varphi : H \mapsto [0,+\infty]$ be a lower semicontinuous convex function non identically $+\infty$.

The following conditions will be assumed:

(i) For every $\varepsilon > 0$, $(I + \varepsilon A)^{-1} D(\varphi) \subset D(\varphi)$ and there exists $C > 0$ such that

$$\varphi((I + \varepsilon A)^{-1}x) \le \varphi(x) + C\varepsilon \quad \text{for all } x \in D(\varphi) .$$

(ii) $D(\varphi) \subset D(A)$ and A^0 is bounded on every level set $\{u \,;\, |u|^2 + \varphi(u) \le M\}$.

Theorem 5: *Assume that conditions* (i), (ii) *are satisfied. Let* x *and* f *be given satisfying*

(5.1) $x \in D(\varphi)$, $f \in L^2(0,T;H)$, $\varphi(f) \in L^1(0,T)$.

Then, for every $\alpha > 0$ *there exists a unique absolutely continuous function* $u : [0,T] \mapsto H$ *such that*

(5.2) $\frac{du}{dt} + Au + \alpha u \ni \alpha f$ a.e. $t \in]0,T[$

(5.3) $u(0) = x$.

Moreover, $\varphi(u) \in L^{\infty}(0,T)$.

P r o o f. For every $\varepsilon > 0$ and all positive λ the equation

$$\frac{du}{dt} + A_\lambda u + \varepsilon\partial\varphi(u) + \alpha\, u \ni \alpha f \quad \text{a.e. } t \in\,]0,T[$$

(5.4)

$$u(0) = x$$

has a unique solution $u_\lambda \in C(0,T;H)$ such that $\frac{du_\lambda}{dt} \in L^2(0,T;H)$. We notice that condition (i) implies that

$$(A_\lambda u,\ w) + C \geq 0 \quad \text{for every} \quad w \in \partial\varphi(u) \ .$$

Thus, by a simple calculation involving (5.4) one obtains the estimate

$$\varphi(u_\lambda(t)) + \int_0^t |\partial\varphi(u_\lambda)|^2\, ds + \alpha \int_0^t \varphi(u_\lambda) ds \leq \alpha \int_0^t \varphi(f) ds + C$$

for every $t \in [0,T]$. By condition (ii) this clearly implies that $|A_\lambda u_\lambda(t)|$ are uniformly bounded on $[0,T]$ and $\frac{du_\lambda}{dt}$ are bounded in $L^2(0,T;H)$. It follows then by a standard device that $u_\lambda(t)$ converges uniformly on $[0,T]$ to the solution u_ε of the equation

$$\frac{du}{dt} + Au + \varepsilon\partial\varphi(u) + \alpha u \ni f \quad \text{a.e. } t \in\,]0,T[$$

(5.5)

$$u(0) = x.$$

Moreover, the estimate

$$(5.6) \qquad \varphi(u_\varepsilon(t)) + \varepsilon \int_0^t |\partial\varphi(u_\varepsilon)|^2 ds \leq \alpha \int_0^t \varphi(f) ds + C$$

holds for every $\varepsilon > 0$ and all $t \in [0,T]$.

Thus $u_\varepsilon(t)$ converges uniformly on $[0,T]$ to $u(t)$. By (5.6) it follows that $\varphi(u(t)) \leq \alpha_0\int^t \varphi(f) ds + C$ for all $t \in [0,T]$. By assumption (ii) this implies that $u(t) \in D(A)$ for all $t \in [0,T]$ and $|A^0 u(t)|$ is bounded on $[0,T]$. It is easy to see that u is a weak solution of the equation (5.2), i.e.,

$$\frac{1}{2}\, |u(t) - \bar{x}|^2 \leq \frac{1}{2} |u(s) - \bar{x}|^2 + \int_s^t (\alpha\, (f(\tau) - u(\tau)) - \bar{y},$$

$$u(\tau) - \bar{x}) d\tau$$

for all $[\bar{x},\bar{y}] \in A$ and $0 < s \leq t < T$.
In particular, it follows that

$$|u(t)-\bar{x}| \leq |u(s)-\bar{x}| + \int_s^t |\alpha(f(\tau)-u(\tau))-\bar{y}|\, d\tau, \quad 0 < s \leq t < T.$$

Taking $\bar{x} = u(s)$ and $\bar{y} = A^o u(s)$ we deduce that

$$|u(t)-u(s)| \leq M(t-s) + \alpha \int_s^t |f(\tau)|\, d\tau\ , \quad 0 < s \leq t < T\ .$$

Hence u is absolutely continuous on [0,T]. In particular, this implies that $\frac{du}{dt}$ exists a.e. on]0,T[and from (5.4) one deduces by a standard argument that $\frac{du}{dt} + Au + \alpha u \ni \alpha f$ a.e. $\in$]0,T[.

Remark. In condition (ii) it suffices to assume that $\varphi((I+\varepsilon A)^{-1}x) \leq \varphi(x) + C(|x|)\varepsilon$ where C is locally bounded on $[0,+\infty[$.

Let $\hat{\varphi} : H \mapsto]-\infty\ ,+\infty]$ be the recession function of φ , i.e.

$$\hat{\varphi}(u) = \lim_{\lambda \to +\infty} \varphi(u_o + \lambda u)/\lambda$$

where u_o is any element of $D(\varphi)$. We recall that

$$\hat{\varphi}(u) = \sup \{(u,v);\ \ v \in D(\varphi^*)\}$$

where $\varphi^*(u) = \sup \{(u,x) - \varphi(x)\}$ is the conjugate function of φ (see e.g. J. MOREAU [15]). The following Corollary is an immediate consequence of Theorem 5.

Corollary 3: *Let* A *satisfy conditions* (i), (ii). *Let* $x \in D(\varphi)$ *and let* $f \in L^2(0,T;H)$ *be such that*

$$\hat{\varphi}(f) \in L^1(0,T).$$

Then there exists a unique absolutely continuous function $u : [0,T] \mapsto H$ *such that*

$$(5.7)\qquad \begin{aligned} &\frac{du}{dt} + Au \ni f \quad \text{a.e. } t \in]0,T[\\ &u(0) = x. \end{aligned}$$

In addition, $\varphi(u) \in L^\infty(0,T)$ and $\frac{du}{dt} - f \in L^\infty(0,T;H)$.

Let K be a closed convex subset of H and let Q_K be the asymptotic cone of K, i.e. $Q_K = \{x \in H;\ x + K \subset K\}$.

Thus by taking $\varphi = I_K$ the indicator function of K we get

Corollary 4. *Let* A *be maximal monotone in* H x H *and let* K *be a closed convex subset of* H.

Assume

(a) $K \subset D(A)$ *and* A^0 *is bounded on every bounded subset of* K.

(b) $(I+\varepsilon A)^{-1}\ K \subset K$ *for all* $\varepsilon > 0$.

Then for every $x \in K$ *and for every* $f \in L^2(0,T;H)$ *satisfying*

(5.8) $f(t) \in Q_K$ a.e. $t \in]0,T[$

the equation (5.7) *has a unique absolutely continuous solution* u *satisfying*

$$\frac{du}{dt} - f \in L^\infty(0,T;H), \quad u(t) \in K \quad \text{a.e. } t \in]0,T[.$$

We finally observe that conditions (i), (ii) hold in particular if there exists a continuous monotone increasing function $\omega : \mathbf{R}^+ \mapsto \mathbf{R}^+$ such that $u \mapsto \omega(|A^0u|)$ is convex on H and $\lim_{r\to\infty} \omega(r) = \infty$.

REFERENCES

[1] V. Barbu, A class of boundary problems for second order abstract differential equations, J. Fac. Sci. Univ. Tokyo, 19, p. 295-319 (1972).

[2] V. Barbu, Sur une problème aux limites pour une class d'equations differentielles du deuxième ordre en t. C.R. Acad. Sc. Paris, t. 274, p. 459-462 (1972).

[3] V. Barbu, Regularity properties of some nonlinear evolution equations, Revue Roumaine Math. Pure et Appl. (to appear).

[4] H. Brezis, Propriétés régularisantes de certain semigroupes non lineaires, Israel J. Math. Vol. 9, 4, p. 513-534 (1971).

[5] H. Brezis, Equations d'evolution du second ordre associées à des opérateurs monotones, Israel J. Math. 12, p. 51-60 (1972).

[6] H. Brezis, Opérateurs maximaux monotones et semi-grupes de contractions dans les espaces de Hilbert. Math. Studies, 5, North Holland, 1973.

[7] H. Brezis, Problemes unilatéraux, J. Math. Pures Appl. 51, 1-164, (1972).

[8] H. Brezis, Monotonicity methods in Hilbert spaces and some applications to nonlinear diff. equation, Contributions to Nonlinear Functional Analysis E. Zarnatonello ed. Acad. Press (1971) p. 101-156.

[9] M.Crandall and A.Pazy, Semigroups of nonlinear contractions and dissipative sets, J. Funct. Analysis, 3, p. 376-418 (1969).

[10] T. Kato, Accretive operators and nonlinear evolution equations in Banach spaces. Proc. Symp. Pure Math. vol. 14, F. Browder ed. Amer. Math. Soc. p. 138-161 (1970).

[11] Y. Konishi, Compacité de resolvantes des opérateurs maximaux cycliquement monotones. Proc. Japan Academy, Vol. 49, p. 303-305 (1973).

[12] Y. Komura, Differentiability of nonlinear semigroups, J. Math. Soc. Japan 21, p. 375-402 (1972).

[13] S.G. Krein, Linear Differential Equations in Banach Spaces, Nauka, Moskva (1967).

[14] J.L. Lions, Quelques methodes de resolution des problèmes aux limites non-lineaires. Dunod et Gauthier-Villars (1969).

[15] J. Moreau, Fonctionnelles Convexes, College de France 1966-1967.

[16] R.T. Rockafellar, Convex functions, monotone operators and variational inequalities, Theory and Applications of Monotone Operators" A. Ghizetti Ed. p. 35-66, Oderisi, Gubio 1969.

Faculty of Mathematics, University of Iaşi, Romania

CLASSES D'INTERPOLATION ASSOCIÉES À UN OPÉRATEUR MONOTONE ET APPLICATIONS

Haim Brezis
Paris (France)

Nous présentons ici un travail récent de D. BREZIS (cf. [1], [2] et rédaction détaillée à paraitre). Etant donné un opérateur maximal monotone A, de domaine D(A), on introduit des classes d'interpolation intermédiaires entre D(A) et $\overline{D(A)}$ qui généralisent les espaces usuels d'interpolation de la théorie linéaire (cf. les travaux de J.L. LIONS, J. PEETRE etc.). On établit l'équivalence de diverses formulations possibles.
Cette étude est motivée par le probleme suivant (soulevé par J.L. LIONS). On considère l'équation parabolique non linéaire

$$\frac{\partial u}{\partial t} - \sum_i \frac{\partial}{\partial x_i}\left(\left|\frac{\partial u}{\partial x_i}\right|^{p-2} \frac{\partial u}{\partial x_i}\right) = 0 \quad \text{sur} \quad \Omega \times]0,+\infty[$$

$$u(x,t) = 0 \quad \text{sur} \quad \partial\Omega \times]0,+\infty[$$

$$u(x,0) = u_o(x) \quad \text{sur} \quad \Omega$$

On sait que si $u_o \in L^2(\Omega)$, alors

$$\lim_{t \to o} |u(.,t) - u_o(.)|_{L^2} = 0$$

De plus si $u_o \in W_0^{1,\infty}(\Omega) \cap W^{2,2}(\Omega)$ (par exemple!) alors

$$|u(.,t) - u_o(.)|_{L^2} \leq Ct .$$

Quelle estimation de $|u(.,t) - u_o(.)|_{L^2}$ peut-on obtenir lorsque la donnée initiale u_o est très régulière ($C^\infty(\overline{\Omega})$ par exemple), mais $u_o \neq 0$ sur $\partial\Omega$? (autrement dit la donnée initiale ne vérifie pas la condition aux limitex). Nous montrerons au § II que l'on a alors une estimation de la forme

$$|u(.,t) - u_o(.)| \leq Ct^{1/2p} .$$

1. CLASSES D'INTERPOLATION ASSOCIÉES À UN OPÉRATEUR MONOTONE

1.1 CAS GÉNÉRAL

Soit H un espace de Hilbert sur **R** et soit A un opérateur maximal monotone (multivoque) de H (cf. par exemple [3]), de domaine D(A).

On pose

$$J_\lambda = (I + \lambda A)^{-1} \quad (\lambda > 0) \quad \text{et} \quad A_\lambda = \frac{1}{\lambda}(I - J_\lambda) . \text{ 1)}$$

On définit pour $0 < \alpha < 1$ et $1 \leq p \leq \infty$

$$\mathcal{B}_{\alpha,p} = \left\{ x \in \overline{D(A)} ; \quad \frac{|x - J_t x|}{t^\alpha} \in L^p_*(0,1) \right\} \text{ où } L^p_*(0,1) = \{ f : [0,1] \mapsto \mathbf{R} ;$$

$$f \text{ mesurable et } \int_0^1 |f(t)|^p \frac{dt}{t} < +\infty \} . \text{ 2)}$$

On prouve aisément que:

- si $\alpha > \alpha'$ alors $\mathcal{B}_{\alpha,p} \subset \mathcal{B}_{\alpha',q}$ pour tout p et q
- si $p \leq q$ alors $\mathcal{B}_{\alpha,p} \subset \mathcal{B}_{\alpha,q}$ pour tout α.

Le lemme suivant sera utile par la suite.

---

1) Cf. la notation p. 15 (reviewers' remark).

2) $L^\infty_*(0,1) = \{ f : [0,1] \mapsto \mathbf{R} ;$ f mesurable et $\sup\text{ess}_{t \in [0,1]} |f(t)| < +\infty \}$. (reviewers' remark).

Lemme 1. *Pour tout* $u_o \in \overline{D(A)}$, *la fonction* $t \mapsto J_t u_o$ *est lipschitzienne sur* $[\delta, +\infty[$ *pour tout* $\delta > 0$ *et p.p. sur* $[0, +\infty[$ *on a*

$$\left|\frac{d}{dt} J_t u_o\right| \le |A_t u_o| \le |A^o v| + \frac{1}{t} |u_o - v|$$

pour tout $v \in D(A)$.[3]

D é m o n s t r a t i o n: La 2ème inégalité est immédiate:

$$|A_t u_o| \le |A_t u_o - A_t v| + |A_t v| \le \frac{1}{t} |u_o - v| + |A^o v| .$$

Pour la 1ère inégalité remarquons que

$$(AJ_{t+h} u_o - AJ_t u_o , J_{t+h} u_o - J_t u_o) \ge 0 ;$$

ce qui conduit à

$$\left(\frac{u_o - J_{t+h} u_o}{t + h} - \frac{u_o - J_t u_o}{t}, J_{t+h} u_o - J_t u_o\right) \ge 0$$

et par suite

$$\frac{1}{t + h} |J_{t+h} u_o - J_t u_o|^2 \le \frac{h}{t(t+h)} |J_{t+h} u_o - J_t u_o| \, |J_t u_o - u_o| .$$

Theorème 1. *Soit* u_o $\overline{D(A)}$; *alors* $u_o \in \mathscr{B}_{a,p}$ *si et seulement si*

$$(1) \quad t^{1-a} \left|\frac{d}{dt} J_t u_o\right| \in L^p_*(0,1) .$$

D é m o n s t r a t i o n: Comme $\left|\frac{d}{dt} J_t u_o\right| \le \frac{|u_o - J_t u_o|}{t}$ il est immédiat que $u_o \in \mathscr{B}_{a,p} \Longrightarrow (1)$.
Inversement

$$|u_o - J_t u_o| \le \int_0^t \left|\frac{d}{dt} J_t u_o\right| ds .$$

On en déduit que $(1) \Longrightarrow u_o \in \mathscr{B}_{a,p}$ grâce à l'inégalité de Hardy

---

[3] Pour la définition de A^o voir p. 46 (reviewers' remark).

$$(2)\qquad \left\| \frac{1}{t^{\alpha}} \int_0^t \varphi(s)\, \frac{ds}{s} \right\|_{L^p_*} \leq \frac{1}{\alpha} \left\| \frac{1}{t^{\alpha}}\, \varphi(t) \right\|_{L^p_*}$$

(pour toute fonction φ).

Theorème 2. (*methode des traces*). *Soit* $u_0 \in \overline{D(A)}$; *alors* $u_0 \in \mathcal{B}_{\alpha,p}$ *si et seulement si*

(3) *il existe une fonction* $v(t)$ *absolument continue sur* $]0,1[$, *continue sur* $[0,1]$ *telle que* $v(0) = u_0$, $v(t) \in D(A)$ p.p. $t^{1-\alpha}|A^o v(t)| \in L^p_*(0,1)$ *et* $t^{1-\alpha}\left|\frac{dv}{dt}(t)\right| \in L^p_*(0,1)$.

D é m o n s t r a t i o n: $u_0 \in \mathcal{B}_{\alpha,p} \Rightarrow (3)$: il suffit de choisir $v(t) = J_t u_0$.

Inversement, on a grâce au Lemme 1

$$\left|\frac{d}{dt} J_t u_0\right| \leq \frac{1}{t}\, |u_0 - v(t)| + |A^o v(t)|$$

$$\leq \frac{1}{t} \int_0^t \left|\frac{dv}{dt}(s)\right| ds + |A^o v(t)|$$

On conclut à l'aide de l'inégalité de Hardy et du Théorème 1.

Par une méthode tout à fait similaire on démontre le

Theorème 2'. *Soit* $u_0 \in \overline{D(A)}$; *alors* $u_0 \in \mathcal{B}_{\alpha,p}$ *si et seulement si*

(4) *il existe deux fonctions mesurables* $v_1(t)$ et $v_2(t)$ *telles que* $u_0 = v_1(t) + v_2(t)$, $v_1(t) \in D(A)$ p.p., $t^{1-\alpha}|A^o v_1(t)| \in L^p_*(0,1)$ et $t^{-\alpha}|v_2(t)| \in L^p_*(0,1)$.

Nous sommes maintenant en mesure d'en déduire un théorème d'interpolation.

Corollaire 1. *Soit* A_1 (*resp.* A_2) *un opérateur maximal monotone sur* H_1 (*resp.* H_2).
Soit T *une application de* $D(A_1)$ *dans* $D(A_2)$ *et de* $\overline{D(A_1)}$ *dans* $\overline{D(A_2)}$ *telle que*

$$\forall \ x, y \in \overline{D(A_1)} \qquad |Tx - Ty| \leq L|x - y|$$

$$\forall \ x \in D(A_1) \qquad |A_2 Tx| \leq |A_1 x| + \omega(|x|)$$

(ω *continu*).
Alors T *applique* $\mathcal{B}_{\alpha,p}(A_1)$ *dans* $\mathcal{B}_{\alpha,p}(A_2)$.

D é m o n s t r a t i o n: Soit $v_1(t) = (I + tA_1)^{-1}u_o$. On a alors

$$T\,u_o = T\,v_1(t) + (T\,u_o - T\,v_1(t))$$

avec

$$t^{1-\alpha}|A_2 T v_1(t)| \leq t^{1-\alpha}|A_1 v_1(t)| + t^{1-\alpha} \ (|v_1(t)|) \ ,$$

$$t^{-\alpha}|Tu_o - Tv_1(t)| \leq Lt^{-\alpha}|u_o - J_t \ A_1 u_o| \ .$$

On conclut à l'aide du Théorème 2' que $Tu_o \in \mathcal{B}_{\alpha,p}\ (A_2)$ dès que $u_o \in \mathcal{B}_{\alpha,p}\ (A_1)$.

Theorème 3 (**méthode** K). *Pour* $u_o \in \overline{D(A)}$ *on pose*

$$K(t,u_o) = \operatorname*{Inf}_{v \in D(A)} \{|u_o - v| + t|A^o v|\} \ .$$

Alors $u_o \in \mathcal{B}_{\alpha,p}$ *si et seulement si*

$$(5) \qquad t^{-\alpha}\, K(t,u_o) \in L^p_* \ (0,1) \ .$$

D é m o n s t r a t i o n: Prenant $v = J_t u_o$, il est clair que $K(t,u_o) \leq 2|u_o - J_t u_o|$ et donc $u_o \in \mathcal{B}_{\alpha,p} \Longrightarrow (5)$.

Inversement, d'après le Lemme 1

$$\left|\frac{d}{dt}\ J_t u_o\right| \leq \frac{1}{t}\ K(t,u_o)$$

et on conclut à l'aide du Théorème 1 que (5) $\Longrightarrow u_o \in \mathcal{B}_{\alpha,p}$.

La caractérisation suivante de $\mathcal{B}_{\alpha,p}$ fait intervenir le semi-

-groupe $S(t)u_o$ engendré par $-A$. "En gros" $u(t) = S(t)u_o$ représente la solution de l'équation

$$\begin{cases} \frac{du}{dt} + Au \ni 0 \quad \text{sur} \quad [0, +\infty[\\ u(0) = u_o \end{cases}$$

Pour les détails cf. [3].

Theorème 4. *Soit* $u_o \in \overline{D(A)}$; *alors* $u_o \in \mathcal{B}_{\alpha,p}$ *si et seulement si*

$$(6) \qquad \left| \frac{u_o - S(t)u_o}{t^\alpha} \right| \in L^p_* (0,1) .$$

De plus on a

$$\frac{1}{3} \left\| \frac{u_o - S(t)u_o}{t^\alpha} \right\|_{L^p_*} \le \left\| \frac{u_o - J_t u_o}{t^\alpha} \right\|_{L^p_*} \le 6 \left\| \frac{u_o - S(t)u_o}{t^\alpha} \right\|_{L^p_*} .$$

La $1^{\text{ère}}$ inégalité est facile et résulte du lemme qui suit; la $2^{\text{ème}}$ inégalité est très délicate à démontrer et nous renvoyons le lecteur à la rédaction détaillée de D. BREZIS.

Lemme 2. *Pour tout* $u_o \in \overline{D(A)}$, *on a*
$|u_o - S(t)u_o| \le 3 \, |u_o - J_t u_o|$.

D é m o n s t r a t i o n: On a pour tout $v \in D(A)$, $|u_o - S(t)u_o| \le |u_o - v| + |v - S(t)v| + |S(t)v - S(t)u_o| \le 2\,|u_o - v| + |v - S(t)v| \le 2|u_o - v| + t\,|A^o v|$.

Prenant en particulier $v = J_t u_o$, on obtient le résultat.

1.2 CAS OÙ $A = \partial\varphi$

Soit $\varphi : H \mapsto]-\infty, +\infty]$ une fonction convexe s.c.i., $\varphi \not\equiv +\infty$. On pose

$$D(\varphi) = \{u \in H; \ \varphi(u) < +\infty\}$$

et

$$\partial\varphi(u) = \{f \in H;\ \varphi(v) - \varphi(u) \geq (f, v - u)\ \ \forall v \in H\}.$$

On sait (cf. par exemple [3]) que $\partial\varphi$ est maximal monotone et que, de plus, le semigroupe S(t) engendré par $-\partial\varphi$ a un effet régularisant i.e.

$$\forall\, t > 0\quad S(t)\,\overline{D(A)} \subset D(A) \quad \text{et}$$

$$|A^{o}\, S(t)u_{o}| \leq |A^{o}v| + \frac{1}{t}\,|u_{o} - v| \quad \forall\, v \in D(A).$$

On a aussi les

Lemme 3. *Pour tout* $u_{o} \in \overline{D(A)}$ *on a*

$$|A^{o}\, S(t)u_{o}| \leq \frac{1}{t}\,|u_{o} - S(t)u_{o}| \quad \forall\, t \in {>}0.$$

D é m o n s t r a t i o n: On multiplie scalairement l'équation

$$\frac{du}{dt} + \partial\varphi(u) \ni 0 \quad \text{par} \quad t\,\frac{du}{dt}(t) \quad \text{où} \quad u(t) = S(t)u_{o}.$$

Il vient donc

$$t\,\left|\frac{du}{dt}(t)\right|^{2} + t\,\frac{d}{dt}\,\varphi(u(t)) = 0.$$

Après intégration sur $]0,T[$, on obtient

$$\int_{0}^{T} t\,\left|\frac{du}{dt}(t)\right|^{2} dt + T\,\varphi(u(T)) \leq \int_{0}^{T} \varphi\,(u(s))\,ds.$$

Or

$$\varphi(v) - \varphi(u(t)) \geq -\left(\frac{du}{dt}(t),\ v-u(t)\right)$$

et par suite

$$\int_{0}^{T} \varphi(u(t))dt \leq T\,\varphi(v) - \frac{1}{2}\,|u(T) - v|^{2} + \frac{1}{2}\,|u_{o} - v|^{2}.$$

D' où

$$\int_{0}^{T} t\,\left|\frac{du}{dt}(t)\right|^{2} dt + T\,\varphi(u(T) \leq T\,\varphi(v) + \frac{1}{2}\,|u_{o} - v|^{2}.$$

Prenant $v = u(T)$ et utilisant le fait que $t \mapsto \left|\frac{du}{dt}(t)\right|$ est décurissant on arrive à $\frac{1}{2} T^2 \left|\frac{du}{dt}(T)\right|^2 \le \frac{1}{2} |u_o - u(T)|^2$.

Lemme 4. *Pour tout* $u_o \in \overline{D(A)}$ *on a*

$$|u_o - J_t u_o| \le 3\, |u_o - S(t)u_o| \quad \forall\, t \ge 0 .$$

D é m o n s t r a t i o n: On a

$$|u_o - J_t u_o| \le |u_o - v| + |v - J_t v| + |J_t v - J_t u_o|$$

$$\le 2\, |u_o - v| + |t\, A^o v| .$$

Prenant $v = S(t)u_o$ et appliquant le Lemme 3 on obtient le résultat désiré.

Remarque. Le Lemme 4, combiné au Lemme 2 fournit une démonstratior élémentaire du Théorème 4 dans le cas où $A = \partial\varphi$.

Nous sommes maintenant en mesure de prouver le

Theorème 5. *Soit* $u_o \in \overline{D(A)}$; *alors* $u_o \in \mathcal{B}_{\alpha,p}$ *si et seulement si*

$$(7) \quad t^{1-\alpha} \left|\frac{d}{dt} S(t)u_o\right| \in L^p_*(0,1) .$$

D é m o n s t r a t i o n: Comme $\left|\frac{d}{dt} S(t)u_o\right| \le \frac{1}{t} |u_o - S(t)u_o|$ (Lemme 3), il est clair $u_o \in \mathcal{B}_{\alpha,p} \Longrightarrow (7)$.
Inversement, on a

$$|u_o - u(t)| \le \int_0^t \left|\frac{du}{dt}(s)\right| ds ,$$

et l'on conclut, grâce à l'inégalité de Hardy que

$$(7) \Longrightarrow u_o \in \mathcal{B}_{\alpha,p} .$$

Indiquons brièvement d'autres caractérisations de $\mathcal{B}_{\alpha,p}$.

Theorème 6. *On suppose* $0 < \alpha < \frac{1}{2}$ *et* $u_o \in \overline{D(A)}$. *Alors* $u_o \in \mathcal{B}_{\alpha,p}$ *si et seulement si*

(8) $t^{1/2-\alpha} \, |\varphi(S(t)u_0)|^{1/2} \in L_*^p \, (0,1)$

ou bien

(9) $t^{1/2-\alpha} \, |\varphi(J_t u_0)|^{1/2} \in L_*^p \, (0,1)$

ou bien

(10) $t^{1/2-\alpha} \, |\varphi_t(u_0)|^{1/2} \in L_*^p \, (0,1)$

avec $\varphi_t(u_o) = \operatorname*{Inf}_{v \in H} \left\{ \frac{1}{2t} \, |u_0 - v|^2 + \varphi(v) \right\}$.

Corollaire 2. *On suppose* $0 < \alpha < \frac{1}{2}$ *et* $u_0 \in \overline{D(A)}$. *Alors* $u_0 \in \mathcal{B}_{\alpha,p}$ *si et seulement si*

(11) *il exite 2 fonctions mesurables* $v_1(t)$ *et* $v_2(t)$ *telles que* $u_0 = v_1(t) + v_2(t)$ *et* $v_1(t) \in D(\varphi)$ p. p., $t^{1/2-\alpha} \, |\varphi(v_1(t))|^{1/2} \in L_*^p \, (0,1)$ *et* $t^{-\alpha} |v_2(t)| \in L_*^p \, (0,1)$.

Indiquons enfin que si $0 < \alpha < 1 - \frac{1}{p}$ il existe des caractérisations de $\mathcal{B}_{\alpha,p}$ faisant intervenir les dérivées fractionnaires de $S(t)u_0$.

2. EQUATIONS PARABOLIQUES NONLINÉAIRES AVEC DÉFAUT D'AJUSTEMENT

Soit $\Omega \subset \mathbf{R}^n$ un ouvert borné de frontière régulière.

Theorème 7. *Soit* $p \geq 2$; *on considere l'equation*

$$\frac{du}{dt} - \sum_i \frac{d}{dx_i} \left(\left| \frac{du}{dx_i} \right|^{p-2} \frac{du}{dx_i} \right) = 0 \qquad \textit{sur} \quad \Omega \times]0, +\infty[$$

$$u(x,t) = 0 \qquad \textit{sur} \quad \partial\Omega \times]0, +\infty[$$

$$u(x,0) = u_0(x) \quad \textit{sur} \quad \Omega$$

Alors si $u_o \in W^{1,p}(\Omega)$ *on a*

$$|u(.,t) - u_o(.)|_{L^2(\Omega)} \leq C\, t^{1/2p}$$

D é m o n s t r a t i o n: On travaille dans $H = L^2(\Omega)$ avec l'opérateur $Au = -\sum_i \frac{d}{dx_i}\left(\left|\frac{du}{dx_i}\right|^{p-2} \frac{du}{dx_i}\right)$ de domaine

$$D(A) = \{u \in W_o^{1,p}(\Omega);\ Au \in L^2(\Omega) \text{ au sens des distributions}\}.$$

Il est aisé de vérifier que A est maximal monotone et que de plus $A = \partial\varphi$ avec

$$\varphi(u) = \begin{cases} \frac{1}{p}\int_\Omega \sum_i \left|\frac{du}{dx_i}\right|^p dx & \text{pour } u \in W_o^{1,p}(\Omega) \\ +\infty & \text{pour } u \in L^2(\Omega),\ u \notin W_o^{1,p}(\Omega). \end{cases}$$

On a ainsi à prouver que $W^{1,p}(\Omega) \subset \mathscr{B}_{1/2p,\infty}$.
Compte tenu des résultats du § I, il y a diverses possibilités pour établir cette inclusion; indiquons en deux:

Première approche; on utilise la définition de $\mathscr{B}_{\alpha,p}$. On considère donc un problème elliptique de perturbations singulières

$$(12) \quad u_\varepsilon - \varepsilon \sum_i \frac{d}{dx_i}\left(\left|\frac{du_\varepsilon}{dx_i}\right|^{p-2} \frac{du_\varepsilon}{dx_i}\right) = u_o \quad \text{sur } \Omega$$

$$u_\varepsilon = 0 \quad \text{sur } \partial\Omega$$

et il s'agit d'établir l'estimation

$$|u_o - u_\varepsilon|_{L^2} \leq C\, \varepsilon^{1/2p}.$$

Pour ce faire, on multiplie (12) d'abord par u_ε. Il vient

$$(13) \quad |u_\varepsilon|_{L^2} \leq C$$

$$(14) \quad \varepsilon \|u_\varepsilon\|^p_{W^{1,p}} \leq C$$

Ensuite on multiplie (12) par $u_\varepsilon - u_o$ et on obtient

$$|u_\varepsilon - u_o|^2_{L^2} + \varepsilon \, \|u_\varepsilon\|^p_{W^{1,p}} \leq \varepsilon \, \|u_o\|_{W^{1,p}} \|u_\varepsilon\|^{p-1}_{W^{1,p}} +$$

$$+ \varepsilon \int_\Omega \sum_i \left|\frac{du_\varepsilon}{dx_i}\right|^{p-2} \frac{du_\varepsilon}{dx_i} \cos(n,x_i) \, u_o d\sigma .$$

Par suite

$$|u_\varepsilon - u_o|^2_{L^2} \leq C \, \varepsilon^{1/p} + \varepsilon \int_{d\Omega} \sum_i \left|\frac{du_\varepsilon}{dn}\right|^{p-1} |\cos(n,x_i)|^p |u_o| \, d\sigma .$$

Comme $u_o \in L^p(\partial\Omega)$, il suffit donc de vérifier

$$(15) \qquad \varepsilon \int_{d\Omega} \left|\frac{du_\varepsilon}{dn}\right|^p \sum_i |\cos(n,x_i)|^p \, d\sigma \leq C .$$

L'estimation (15) s'obtient en multipliant (12) par $\sum_j \alpha_j(x) \frac{du_\varepsilon}{dx_j}(x)$ où (α_j) est un champ de vecteurs, régulier sur Ω et normal à $\partial\Omega$ sur $\partial\Omega$.

Deuxième approche: On utilise, par exemple, le Corollaire 2. Par cartes locales on se ramène au cas où Ω est un demi-espace. On peut alors choisir

$$v_1(x_1,x_2,\ldots,x_n,t) = u_o(x_1,x_2,\ldots,x_n) - u_o(x_1,\ldots,x_{n-1},t^{-1/p}x_n),$$

$$v_2(x_1,x_2,\ldots,x_n,t) = u_o(x_1,x_2,\ldots,x_{n-1},t^{-1/p}x_n).$$

On vérifie aisément que $\forall \, t > 0$, $v_1(t) \in D(\varphi)$ et

$$\varphi(v_1(t)) \leq \|u_o\|^p_{W^{1,p}} \, t^{1/p-1}$$

de sorte que $t^{1/2-1/2p} \, |\varphi(v_1(t))|^{1/p} \in L^\infty(0,1)$.
Par ailleurs

$$|v_2(t)|_{L^2} \leq t^{1/2p} \, |u_o|_{L^2} .$$

Une technique similaire permet d'établir les résultats suivants:

Theorème 8. *On considère l'équation*

$$\begin{cases} \frac{du}{dt} - \Delta u = 0 & \textit{sur} \quad \Omega \times]0, +\infty[\\ -\frac{du}{dn} \in \beta(u) & \textit{sur} \quad \partial\Omega \times]0, +\infty[\\ u(x,0)=u_o(x) & \textit{sur} \quad \Omega \end{cases}$$

où β *est un graphe maximal monotone de* $\mathbf{R} \times \mathbf{R}$. *Alors si* $u_o \in H^1(\Omega)$ *on a*

$$|u(.,t) - u_o(.)|_{L^2} \leq Ct^{1/4} .$$

Theorème 9. (Ioss). *On considère le système suivant*

$$\begin{aligned} \frac{du}{dt} - \Delta u &= \text{grad } p & \textit{sur} \quad & \Omega \times]0, +\infty[\\ \text{div } u &= 0 & \textit{sur} \quad & \Omega \times]0, +\infty[\\ u \cdot n &= 0 & \textit{sur} \quad & \partial\Omega \times]0, +\infty[\\ u(x,0) &= u_o(x) & \textit{sur} \quad & \Omega . \end{aligned}$$

Alors si $u_o \in H^1(\Omega)^n$ *avec* div $u_o = 0$ *sur* Ω *et* $u_o.n = 0$ *sur* $\partial\Omega$ *on a*

$$|u(.,t) - u_o(.)|_{L^2} \leq C\, t^{1/4} .$$

RÉFÉRENCES

[1] D. Brezis, Classes d'interpolation associées à un opérateur monotone, C.R. Acad. Sci. 276 (1973), p. 1553-1556.

[2] D. Brezis, Perturbations singulières et problèmes d'évolution avec défaut d'ajustement, C.R. Acad. Sci. 276 (1973), p. 1597-1600.

[3] H. Brezis, Opérateurs maximaux monotones et semi-groupes de contractions, Math. Studies Vol. 5, North-Holland (1973).

Departement des Mathématiques, Université de Paris VI, 9 Quai St. Bernard, Paris, France.

ON INVERSE PROBLEMS FOR k-DIMENSIONAL POTENTIALS

Siegfried Dümmel
Karl-Marx-Stadt (GDR)

1. INTRODUCTION

Let $\mathbf{R}^n$ $(n = 3, 4, 5, \ldots)$ be the n-dimensional Euclidean space, $r(x,y)$ the distance from $x \in \mathbf{R}^n$ to $y \in \mathbf{R}^n$, k an integer with $1 \leq k \leq n-1$, $F \subset \mathbf{R}^n$ a bounded k-dimensional manifold (the assumptions on F will be fixed later), φ a mass distribution on F, i.e., φ is a finite signed measure, defined on a σ-algebra of subsets of F which contains all Borel subsets of F. Then the k-dimensional potential

$$(1.1) \qquad u(x) = \int_F r^{2-n}(x,y)\, d\varphi$$

is a harmonic function in $\mathbf{R}^n - F$.

In this paper we consider the following inverse problem: Let u(x) be a real function harmonic in $\mathbf{R}^n - F$. Under what conditions on u(x) does there exist a mass distribution φ on F such that u(x) is representable in the form (1.1) for all $x \in \mathbf{R}^n - F$? For the case that F is the boundary of a bounded region G and u(x) is either given in G or in $\mathbf{R}^n - \overline{G}$ where $\overline{G}$ is the closure of G one can find such conditions in a paper of G.A. GARRETT [7].

For investigating the problem given above we consider first some properties of the k-dimensional potential (1.1). Then we introduce the concept of fertility of a vector function, and by means of this concept of fertility we can find a solution of the problem considered.

2. k-DIMENSIONAL POTENTIALS

In this section we use the same notations as in Section 1. Further let μ_k be the outer k-dimensional Hausdorff measure in $\mathbf{R}^n$ ($1 \leqq k \leqq n-1$). We assume that the bounded set $F \subset \mathbf{R}^n$ is μ_k-measurable and $\mu_k(F) > 0$. We consider now the k-dimensional potential

$$(2.1) \qquad u_\lambda(x) = \int_F r^{-\lambda}(x,y)\, d\varphi$$

where λ is a real number with $\lambda > 0$ and φ a mass distribution on F. For $\lambda = n-2$ we obtain the potential (1.1). Let for $x_0 \in F$, $\overline{S}_\varrho(x_0)$ ($S_\varrho(x_0)$) be the closed (open) ball with center at x_0 and radius ϱ and let $\overline{\varphi}$ be the total variation of φ. We make the following further assumptions:

a) The limits

$$(2.2) \qquad \delta(x_0) = \lim_{\varrho \to 0} \frac{\varphi(F \cap \overline{S}_\varrho(x_0))}{\mu_k(F \cap \overline{S}_\varrho(x_0))}$$

and

$$(2.3) \qquad \overline{\delta}(x_0) = \lim_{\varrho \to 0} \frac{\overline{\varphi}(F \cap \overline{S}_\varrho(x_0))}{\mu_k(F \cap \overline{S}_\varrho(x_0))}$$

exist and are finite.

b) We have

$$(2.4) \qquad \lim_{\varrho \to 0} \frac{\mu_k(F \cap \overline{S}_\varrho(x_0))}{\Omega_k \varrho^k}$$

where Ω_k is the k-dimensional measure of the k-dimensional unit ball.

c) x_0 is a Lebesgue point of φ.

d) There exists a normal ν on F in x_0.

Under these assumptions on F and φ the potential $u_\lambda(x)$ has the following proporties:

Let $\overline{x}$ be a point of ν and

$$\omega_a = \frac{2\pi^{\frac{a}{2}}}{\Gamma(\frac{a}{2})} \quad (a > 0) .$$

If $\lambda < k$, then

$$\lim_{\overline{x} \to x_0} u_\lambda(\overline{x}) = u_\lambda(x_0) \, .$$

If $\lambda = k$, then

$$\lim_{\overline{x} \to x_0} \frac{u_\lambda(\overline{x})}{\log r^{-1}(\overline{x},x_0)} = \delta(x_0)\,\omega_k \, .$$

If $\lambda > k$, then

$$\lim_{\overline{x} \to x_0} r^{\lambda-k}(\overline{x},x_0)\, u_\lambda(\overline{x}) = \frac{\delta(x_0)\,\omega_\lambda}{\omega_{\lambda-k}} \, .$$

For the definitions of the Lebesgue point of φ and the normal of F, for a discussion of the assumptions a) to d) and for the proofs of the formulated statements compare S. DÜMMEL [1], [5].

Relations (2.3) and (2.4) imply the existence of two numbers $C > 0$ and $\varrho_0 > 0$ such that

$$(2.5) \qquad \overline{\varphi}(F \cap S_\varrho(x_0)) \leq C\varrho^k$$

for all ϱ with $0 < \varrho \leq \varrho_0$. If we use only the condition (2.5) we are also able to make some statements on the behaviour of $u_\lambda(x)$. For this reason we give the following definitions: We say that φ satisfies at $x_0 \in \mathbf{R}^n$ a Hölder condition with the exponent $\gamma > 0$, if there exist two numbers $C > 0$, $\varrho_0 > 0$, such that

$$\overline{\varphi}(F \cap S_\varrho(x_0)) \leqq C\varrho^\gamma$$

for all ϱ with $0 < \varrho \leqq \varrho_0$.

We say that φ satisfies in the set $A \subseteq \mathbf{R}^n$ a uniform Hölder condition with the exponent γ if φ satisfies at every point $x \in A$ a Hölder condition with the exponent γ and the numbers C and ϱ_0 are independent of x.

Instead of the assumptions a) to d) we assume now:

e) φ satisfies in a neighborhood of F a uniform Hölder condition with the exponent $\gamma = k$.

Under the assumption e) on φ the potential $u_\lambda(x)$ has the following properties:

Let $R(x) = r(x,F)$.

If $\lambda < k$, then $u_\lambda(x)$ is continuous in $\mathbf{R}^n$.

If $\lambda = k$, then $\dfrac{u_\lambda(x)}{\log R^{-1}(x)}$ is bounded in $\mathbf{R}^n - F$.

If $\lambda > k$, then $R^{\lambda-k}(x)\, u_\lambda(x)$ is bounded in $\mathbf{R}^n - F$.

One can obtain these properties by using some results of S. DÜMMEL [2], [4].

3. THE FERTILITY OF A VECTOR FUNCTION

Let $\mathfrak{w}(x)$ be a continuous vector function defined in $\mathbf{R}^n$, A a bounded region of $\mathbf{R}^n$, ∂A the boundary of A (sufficiently regular), $\mathfrak{n}$ the outer normal vector of A, $d\sigma$ the surface element of ∂A; then the fertility (in german: Ergiebigkeit) of $\mathfrak{w}$ in A is classically defined to be

$$\eta(A,\mathfrak{w}) = \int_{\partial A} \mathfrak{w}\,\mathfrak{n}\, d\sigma \, .$$

For our investigation it is necessary to have a more general concept of fertility. Such a more general definition was given by W. RINOW [8], who defines first the interval function

$$(3.1) \qquad \alpha(J,\mathfrak{w}) = \int_{\partial J} \mathfrak{w}\,\mathfrak{n}\, d\sigma$$

on the system of all closed intervals J of $\mathbf{R}^n$. The only assumption made on $\mathfrak{w}(x)$ is that in (3.1) the integrals always exist. This interval function is extended to a signed measure η which is called the fertility of $\mathfrak{w}$. It is also possible to define the fertility of $\mathfrak{w}$ as a distribution (compare S. DÜMMEL [6]). We will choose here this latter way.

Let $\mathfrak{D}$ be the set of all real functions defined in $\mathbf{R}^n$ which have continuous partial derivatives of every order and compact support. Let $\mu_i (1 \leqq i \leqq n)$ be the i-dimensional Hausdorff measure (μ_n Lebesgue measure) and $\mathfrak{w}(x)$ a vector function locally integrable on $\mathbf{R}^n$. If there exists a signed measure $\eta(\mathfrak{w})$ defined on the system of all bounded Borel sets of $\mathbf{R}^n$ such that

$$(3.2) \qquad \int_{\mathbf{R}^n} g(x)\, d\eta = - \int_{\mathbf{R}^n} \mathfrak{w}(x) \operatorname{grad} g(x)\, d\mu_n$$

for all $g \in \mathcal{D}$, then $\eta(\mathfrak{w})$ is called the fertility of $\mathfrak{w}$.

Let $K \subset \mathbf{R}^n$ be compact. We denote by $\mathcal{D}_k$ the set of all $g \in \mathcal{D}$ with support in K. Then we have the following proposition:

$\eta(\mathfrak{w})$ exists, is uniquely determined by (3.2) and is σ-finite, iff to every compact set $K \subset \mathbf{R}^n$ there corresponds a positive number a such that

$$(3.3) \qquad \left| \int_K \mathfrak{w}(x) \operatorname{grad} g(x)\, d\mu_n \right| \leqq a \max_{x \in K} |g(x)|$$

for all $g \in \mathcal{D}_K$ (compare L. SCHWARTZ [9] p. 25).

Now we formulate some properties of the fertility. The proofs of the following assertions which are not given here can be found in S. DÜMMEL [6]. Assuming that η exists and is σ-finite we obtain:

1) Let A be a bounded Borel set and α, β real numbers, then

$$\eta(A, \alpha \mathfrak{w}_1 + \beta \mathfrak{w}_2) = \alpha \eta(A, \mathfrak{w}_1) + \beta \eta(A, \mathfrak{w}_2) .$$

2) Let B be a compact subset of $\mathbf{R}^n$ and $\mathfrak{w}(x) = 0$ for all $x \in \mathbf{R}^n - B$, then $\eta(B, \mathfrak{w}) = 0$.

3) Let G be an open bounded subset of $\mathbf{R}^n$ and $\mathfrak{w} \in C^1(G)$, then

$$\eta(G, \mathfrak{w}) = \int_G \operatorname{div} \mathfrak{w}\, d\mu_n .$$

4) Let F be a bounded (n-1)-dimensional oriented manifold of the class C^2, $\mathfrak{n}(x_0)$ the normal vector of F at $x_0 \in F$,

$$g^+(x_0) = \{x \in \mathbf{R}^n \mid x = x_0 + t\mathfrak{n}(x_0),\ t > 0\} ,$$

$$g^-(x_0) = \{x \in \mathbf{R}^n \mid x = x_0 + t\mathfrak{n}(x_0),\ t < 0\} ,$$

$U(F)$ a neighborhood of F. Further let $w \in C^1(U(F) - F)$. For every $x_0 \in F$ let the limits

$$w^+(x_0) = \lim_{\substack{x \to x_0 \\ x \in g^+(x_0)}} w(x) \quad \text{and} \quad w^-(x) = \lim_{\substack{x \to x_0 \\ x \in g^-(x_0)}} w(x)$$

exist. Let w^+ and w^- be μ_{n-1}-integrable on F. Then

$$\eta(F, w) = \int_F (w^+ - w^-)\, \mathfrak{n}\, d\mu_{n-1} .$$

5) Let G be a bounded open set with the boundary ∂G of the class C^2, $\mathfrak{n}(x)$ the outer normal vector of ∂G, $U(\partial G)$ a neighborhood of ∂G, $w \in C^1(U(\partial G))$, then

$$\eta(G, w) = \int_{\partial G} w\, \mathfrak{n}\, d\mu_{n-1} .$$

P r o o f. Let

$$w^*(x) = \begin{cases} w(x) & \text{for } x \in \overline{G} , \\ 0 & \text{for } x \in \mathbf{R}^n - \overline{G} . \end{cases}$$

Then by 2) we have

$$\eta(\overline{G}, w^*) = 0 .$$

On the other hand, it is

$$\eta(\overline{G}, w^*) = \eta(G, w^*) + \eta(\partial G, w^*) = \eta(G, w) + \eta(\partial G, w^*) ,$$

and consequently

$$\eta(G, w) = - \eta(\partial G, w^*) .$$

By 4) we obtain

$$\eta(\partial G, w^*) = - \int_{\partial G} w\, \mathfrak{n}\, d\mu_{n-1} .$$

Hence

$$\eta(G, v) = \int_{\partial G} v n \, d\mu_{n-1} .$$

4. REPRESENTATION OF HARMONIC FUNCTIONS BY k-DIMENSIONAL POTENTIALS

We consider now the problem suggested in the introduction. In this section we assume always that F is a bounded k-dimensional manifold ($1 \leqq k \leqq n-1$) of the class C^2. We denote by F_ϱ the union set of all open balls with radius ϱ and center on F. If ϱ is sufficiently small then the boundary ∂F_ϱ of F_ϱ is an (n-1)-dimensional manifold of the class C^2. By $\frac{d}{d\nu}$ we denote the derivative with respect to the outer normal vector of ∂F_ϱ. Before we state the main theorem we prove two lemmas.

Lemma 1. *Let* u(x) *be a real function defined in* $\mathbf{R}^n - F$ *and satisfying the following conditions:*

a) u *is harmonic in* $\mathbf{R}^n - F$.

b) u *and* grad u *are in* $\mathbf{R}^n$ *locally integrable.*

c) *There exist two positive numbers* ϱ^* *and* b^* *such that*

$$\int_{\partial F_\varrho} |u(x)| \, d\mu_{n-1} \leqq b^* \quad \text{and} \quad \int_{\partial F_\varrho} \left|\frac{du(x)}{d\nu}\right| d\mu_{n-1} \leqq b^*$$

for all ϱ *with* $0 < \varrho \leqq \varrho^*$.

Let y *be a fixed point of* $\mathbf{R}_n - F$ *and*

$$v_1(x) = \text{grad } u(x),$$

$$v_2(x) = r^{2-n}(x,y) \text{ grad } u(x),$$

$$v_3(x) = u(x) \text{ grad } r^{2-n}(x,y),$$

$$v_4(x) = v_2(x) - v_3(x).$$

Then the fertilities $\eta(v_j)$ ($j = 1,2,3,4$) *exist and are* σ*-finite.*

P r o o f. We show that the relation (3.3) is satisfied. Let K be a compact subset of $\mathbf{R}^n$, S an open ball with $K \subset S$, $F \subset S$ and $y \in S$, $S_\varrho(y)$ the open ball with center at y and radius ϱ. Then we have for every function $g \in \mathcal{D}_K$

$$(4.1) \quad \int_K w_j \operatorname{grad} g \, d\mu_n = \int_S w_j \operatorname{grad} g \, d\mu_n = \lim_{\varrho \to 0} \int_{S-(F_\varrho \cup S_\varrho(y))} w_j \operatorname{grad} g \, d\mu_n .$$

Let ϱ_0 be a positive number such that $0 < \varrho_0 \leqq \varrho^*$ and $F_{\varrho_0} \cap S_{\varrho_0}(y) = \emptyset$. We assume $0 < \varrho \leqq \varrho_0$ and consider first the last integral of (4.1) for $j = 1$.

$$\int_{S-(F_\varrho \cup S_\varrho(y))} w_1 \operatorname{grad} g \, d\mu_n = \int_{S-(F_\varrho \cup S_\varrho(y))} \operatorname{grad} u \operatorname{grad} g \, d\mu_n =$$

$$= -\int_{S-(F_\varrho \cup S_\varrho(y))} g \operatorname{div} \operatorname{grad} u \, d\mu_n + \int_{\partial F_\varrho \cup \partial S_\varrho(y)} g \frac{du}{d\nu} d\mu_{n-1} =$$

$$= \int_{\partial F_\varrho} g \frac{du}{d\nu} d\mu_{n-1} + \int_{\partial S_\varrho(y)} g \frac{du}{d\nu} d\mu_{n-1} .$$

We define

$$A = \max_{x \in K} |g(x)|$$

and obtain

$$\left| \int_{\partial F_\varrho} g \frac{du}{d\nu} d\mu_{n-1} \right| \leqq A \int_{\partial F_\varrho} \left| \frac{du}{d\nu} \right| d\mu_{n-1} \leqq b^* A .$$

The derivatives $\frac{du}{dx_i}$ $(i = 1,2,\ldots,n)$ are bounded in $\overline{S}_{\varrho_0}$. Therefore there is a positive number b_0 such that

$$(4.2) \quad \left| \frac{du}{d\nu} \right| = g_0$$

for all ∂S_ϱ with $0 < \varrho \leqq \varrho_0$ and we obtain

$$\int_{\partial \overline{S}_\varrho(y)} g \frac{du}{d\nu} d\mu_{n-1} \leqq A b_0 \, \mu_{n-1} (\partial S_\varrho(y)) \leqq b_0 \omega_n \varrho_0^{n-1} A .$$

Hence

$$\left| \int_K w_1 \text{ grad } g \, d\mu_n \right| \leqq a_1 A$$

with $a_1 = b^* + b_o \omega_n \varrho_o^{n-1}$ and the condition (3.3) is satisfied. We consider now the case j = 2:

$$\int_{S-(F_\varrho \cup S_\varrho(y))} w_2 \text{ grad } g \, d\mu_n = \int_{S-(F_\varrho \cup S_\varrho(y))} r^{2-n} \text{ grad } u \text{ grad } g \, d\mu_n =$$

$$= - \int_{S-(F_\varrho \cup S_\varrho(y))} g \text{ div}(r^{2-n} \text{ grad } u) \, d\mu_n + \int_{\partial F_\varrho \cup S_\varrho(y)} r^{2-n} \frac{du}{d\nu} g \, d\mu_{n-1} =$$

$$= - \int_{S-(F_\varrho \cup S_\varrho(y))} g \text{ grad } r^{2-n} \text{ grad } u \, d\mu_n + \int_{\partial F_\varrho} r^{2-n} \frac{du}{d\nu} g \, d\mu_{n-1} +$$

$$+ \int_{\partial S_\varrho(y)} r^{2-n} \frac{du}{d} g \, d\mu_{n-1} .$$

grad r^{2-n} grad u is integrable in S. Therefore

$$b_1 = \int_S |\text{grad } r^{2-n} \text{ grad } u| \, d\mu_n < + \infty$$

and

$$\left| \int_{S-(F_\varrho \cup S_\varrho(y))} g \text{ grad } r^{2-n} \text{ grad } u \, d\mu_n \right| \leqq b_1 A .$$

$r^{2-n}(x,y)$ is bounded in $\overline{F}_{\varrho_o}$:

$$r^{2-n}(x,y) \leqq b_2$$

for all $x \in \overline{F}_{\varrho_o}$. Therefore we obtain

$$\left| \int_{\partial F_\varrho} r^{2-n} \frac{du}{d\nu} g \, d\mu_{n-1} \right| \leqq b_2 A \int_{\partial F_\varrho} \left| \frac{du}{d\nu} \right| d\mu_{n-1} \leqq b_2 b^* A .$$

Let the constant b_o be defined by (4.2). Then

$$\left| \int_{\partial S_\varrho(y)} r^{2-n} \frac{du}{d\nu} g \, d\mu_{n-1} \right| = \varrho^{2-n} \left| \int_{\partial S_\varrho(y)} \frac{du}{d\nu} g \, d\mu_{n-1} \right| \leqq \varrho^{2-n} b_0 A \omega_n \varrho^{n-1} \leqq$$

$$\leqq b_0 \omega_{n-1} \varrho_0 A \, .$$

Hence

$$\left| \int_K \mathfrak{w}_2 \text{ grad } g \, d\mu_n \right| \leqq a_2 A$$

with $a_2 = b_1 + b_2 b^* + b_0 \omega_n \varrho_0$ and the condition (3.3) is satisfied. In an analogous way one can show the existence of $\eta(\mathfrak{w}_3)$. The existence of $\eta(\mathfrak{w}_4)$ is then clear.

Lemma 2. *Let* u(x) *be a real function defined on* $\mathbf{R}^n$ − F *and satisfying the conditions* a), b), c) *of* Lemma 1. *Let* y *be a fixed point of* $\mathbf{R}^n$ − F *and*

$$\mathfrak{w}_2(x) = r^{2-n}(x,y) \text{ grad } u(x) \, .$$

Then there exists a signed measure φ *with support on* F *such that the fertility*

$$\eta(F,\mathfrak{w}_2) = \int_F r^{2-n}(x,y) \, d\varphi$$

and

$$\varphi(F) = \eta(F, \text{ grad } u) \, .$$

P r o o f. $\eta(\mathfrak{w}_2)$ and $\eta(\text{grad } u)$ exist by Lemma 1. We have by property 5) of the fertility

$$\eta(F_\varrho,\mathfrak{w}_2) = \int_{\partial F_\varrho} \mathfrak{w}_{2n} d\mu_{n-1} = \int_{\partial F_\varrho} r^{2-n}(x,y) \frac{du}{d\nu} d\mu_{n-1} \, .$$

If X_ϱ is a Borel subset of ∂F_ϱ then let

$$\psi_\varrho(X_\varrho) = \int_{X_\varrho} \frac{du}{d\nu} d\mu_{n-1} \, ;$$

ψ_ϱ is of bounded variation where the bound is independent of ϱ. $r^{2-n}(x,y)$ is continuous in a neighborhood of F. Thus there exists a mass distribution φ on F and a sequence (ϱ_m) with $\varrho_m \to 0$ such that

$$\lim_{m\to\infty} \int_{\partial F_{\varrho_m}} r^{2-n}(x,y)\, d\psi_{\varrho_m} = \int_F r^{2-n}(x,y)\, d\varphi .$$

On the other hand, we have

$$\eta(F,w_2) = \lim_{\varrho\to 0} \eta(F_\varrho,w_2) = \lim_{\varrho\to 0} \int_{\partial F_\varrho} r^{2-n}(x,y)\, d\psi_\varrho .$$

Thus

$$\eta(F,w_2) = \int_F r^{2-n}(x,y)\, d\varphi .$$

Further we have

$$\eta\,(F, \text{grad } u) = \lim_{\varrho\to 0} \eta(F_\varrho, \text{grad } u) = \lim_{\varrho\to 0} \int_{\partial F_\varrho} \frac{du}{d\nu}\, d\mu_{n-1} =$$

$$= \lim_{\varrho\to 0} \psi_\varrho(\partial F_\varrho) = \varphi(F) ,$$

which proves the lemma.

Now we are able to formulate the above indicated theorem.

Theorem 1. *Let* u(x) *be a real function defined in* $\mathbf{R}^n - F$ *and satisfying the following conditions:*

a) u *is harmonic in* $\mathbf{R}^n - F$,

$$\lim_{|x|\to\infty} u(x) = 0$$

and there are two positive numbers ϱ_0 *and* b_0 *such that*

$$|x|^2\, |\text{grad } u(x)| \leqq b_0$$

for all $x \in \mathbf{R}^n$ *with* $|x| \geqq \varrho_0$.

b) u *and* grad u *are locally integrable in* $\mathbf{R}^n$.

c) *There exist two positive numbers* ϱ^* *and* b^* *such that*

$$\int_{\partial F_\varrho} |u(x)|\, d\mu_{n-1} \leqq b^* \quad \text{and} \quad \int_{\partial F_\varrho} \left|\frac{du(x)}{d\nu}\right| d\mu_{n-1} \leqq b^*$$

for all ϱ *with* $0 < \varrho \leqq \varrho^*$.

d)

$$\lim_{\varrho \to 0} \int_{\partial F_\varrho} u(x) \frac{\partial r^{2-n}(x,y)}{\partial \nu} d\mu_{n-1} = 0$$

for all $y \in \mathbf{R}^n - F$.

Then there exists a signed measure φ *with support on* F *(a mass distribution on* F*) such that*

$$u(y) = -\frac{1}{(n-2)\omega_n} \int_F r^{2-n}(x,y)\, d\varphi$$

for all $y \in \mathbf{R}^n - F$.

For the measure φ *we have*

$$\varphi(F) = \eta(F, \text{grad } u)$$

where η *is the fertility of* grad u.

P r o o f. By Lemma 2 the fertilities $\eta(\mathfrak{w}_j)$ of the vector functions $\mathfrak{w}_j$ $(j = 1,2,3,4)$ introduced in Lemma 2 exist and are σ-finite. First we consider

$$\mathfrak{w}(x) = \mathfrak{w}_4(x) = r^{2-n}(x,y) \text{ grad } u(x) - u(x) \text{ grad } r^{2-n}(x,y) .$$

Let S_ϱ be an open ball with radius ϱ and center at y such that $F \subset S_\varrho$. We define

$$\mathfrak{w}^*(x) = \begin{cases} \mathfrak{w}(x) & \text{for } x \in \overline{S}_\varrho , \\ 0 & \text{for } x \in \mathbf{R}^n - \overline{S}_\varrho . \end{cases}$$

We have

$$(4.3) \quad \eta(\overline{S}_\varrho, \mathfrak{w}^*) = \eta(S_\varrho - (F \cup \{y\}), \mathfrak{w}^*) + \eta(\partial S_\varrho, \mathfrak{w}^*) + \\ + \eta(F, \mathfrak{w}^*) + \eta(\{y\}, \mathfrak{w}^*) .$$

By property 2) of the fertility we obtain

$$\eta(\overline{S}_\varrho, \mathfrak{w}^*) = 0 .$$

From property 3) of the fertility we see that

$$\eta(S_\varrho-(FU\{y\}),\mathfrak{w}^*) = \int\limits_{S_\varrho-(FU\{y\})} \operatorname{div} \mathfrak{w}\, d\mu_n =$$

$$= \int\limits_{S_\varrho-(FU\{y\})} (\operatorname{grad} r^{2-n} \operatorname{grad} u + r^{2-n}\Delta u - \operatorname{grad} r^{2-n} \operatorname{grad} u - u\Delta r^{2-n})\,d\mu_n = 0 .$$

Let $\mathfrak{n}$ be the outer normal vector of ∂S_ϱ. Then we have by property 4) of the fertility

$$\eta(\partial S_\varrho,\mathfrak{w}^*) = -\int\limits_{\partial S_\varrho} \mathfrak{w}\mathfrak{n}\, d\mu_{n-1} = -\int\limits_{\partial S_\varrho} r^{2-n}\frac{\partial u}{\partial \nu}\, d\mu_{n-1} + \int\limits_{\partial S_\varrho} u\frac{\partial r^{2-n}}{\partial \nu} d\mu_{n-1} =$$

$$= -\varrho^{2-n}\int\limits_{\partial S_\varrho}\frac{\partial u}{\partial \nu}\, d\mu_{n-1} + (2-n)\,\varrho^{1-n}\int\limits_{\partial S_\varrho} u\, d\mu_{n-1} .$$

Hence

$$|\eta(\partial S\ ,\mathfrak{w}^*)| \leqq \varrho^{2-n}\omega_n\varrho^{n-1}\max_{x\in\partial S_\varrho}|\operatorname{grad} u(x)| +$$

$$+ (n-2)\,\varrho^{1-n}\,\omega_n\,\varrho^{n-1}\max_{x\in\partial S_\varrho}|u(x)| .$$

Thus by condition a) of the theorem we obtain

$$\lim_{\varrho\to\infty} \eta(\partial S_\varrho,\mathfrak{w}^*) = 0 .$$

At last we have by using property 5) of the fertility

$$\eta(\{y\},\mathfrak{w}^*) = \eta(\{y\},\mathfrak{w}) = \lim_{h\to o}\eta(S_h(y),\mathfrak{v}) = \lim_{h\to o}\int\limits_{\partial S_h(y)} \mathfrak{w}\mathfrak{n}\,\mu_{n-1} =$$

$$= \lim_{h\to o}\left(h^{2-n}\int\limits_{\partial S_h(y)}\frac{\partial u}{\partial \nu}\, d\mu_{n-1} + (n-2)\,h^{1-n}\int\limits_{\partial S_h(y)} u\, d\mu_{n-1}\right) =$$

$$= (n-2)\,\omega_n u(y) .$$

Thus we obtain from equation (4.3)

$$(4.4)\qquad u(y) = -\frac{1}{(n-2)\,\omega_n}\,\eta(F,\mathfrak{w}) .$$

Using the notation

$$\mathfrak{w}_2(x) = r^{2-n}(x,y) \text{ grad } u(x), \; \mathfrak{w}_3(x) = u(x) \text{ grad } r^{2-n}(x,y)$$

we have

$$(4.5) \qquad \eta(F,\mathfrak{w}) = \eta(F,\mathfrak{w}_2) - \eta(F,\mathfrak{w}_3) \, .$$

By Lemma 2 there is a signed measure φ with support on F such that

$$\eta(F,\mathfrak{w}_2) = \int_F r^{2-n}(x,y) \, d\varphi$$

and

$$\varphi(F) = \eta(F, \text{ grad } u) \, .$$

We consider the second term of (4.5) and obtain by using condition d) of the theorem

$$\eta(F,\mathfrak{w}_3) = \lim_{\varrho \to 0} \eta(F_\varrho, \mathfrak{w}_3) = \lim_{\varrho \to 0} \int_{\partial F_\varrho} \mathfrak{w}_3 \, \mathfrak{n} \, d\mu_{n-1}$$

$$= \lim_{\varrho \to 0} \int_{\partial F_\varrho} u \, \frac{\partial r^{2-n}}{\partial \nu} \, d\mu_{n-1} = 0 \, .$$

Thus we see from (4.4) and (4.5) that

$$u(y) = - \frac{1}{(n-2)\,\omega_n} \int_F r^{2-n}(x,y) \, d\varphi$$

and

$$\varphi(F) = \eta(F, \text{ grad } u) \, .$$

One can expect that the conditions of Theorem 1 are also necessary. But here we can only prove the necessity of these conditions under some restrictions on the mass distribution.

Theorem 2. *Let φ be a mass distribution on F which satisfies on a neighborhood of F a uniform Hölder condition with the exponent $\gamma = k$ (compare Section 2). Then*

$$u(x) = - \frac{1}{(n-2)\,\omega_n} \int_F r^{2-n}(x,y) \, d\varphi$$

satisfies conditions a) to d) of Theorem 1.

P r o o f. Conditions a) and b) are clear. We consider now the first condition of c). Let $y \in \mathbf{R}^n - F$ and $R(y) = r(y,F)$. Using the last propositions of Section 2 we obtain for $k < n-2$ that there is a positive number N such that

$$(4.6) \qquad |u(x)| \leqq N\, R^{k-n+2}(x)$$

for all $x \in \mathbf{R}^n - F$. Further, there are two positive numbers N_0 and ϱ_0 such that

$$(4.7) \qquad \mu_{n-1}(\partial F_\varrho) \leqq N_0 \varrho^{n-1-k}$$

for all ϱ with $0 < \varrho \leqq \varrho_0$ (compare S. DÜMMEL [3], p. 316, proof of Theorem 1). Thus (4.6) and (4.7) imply for all these ϱ

$$\int_{\partial F_\varrho} |u(x)|\, d\mu_{n-1} \leqq NN_0\, \varrho^{k-n+2}\, \varrho^{n-1-k} = NN_0 \varrho \, .$$

If $k = n - 2$ we have

$$|u(x)| \leqq N \log R^{-1}(x)$$

for all $x \in \mathbf{R}^n - F$ and thus by (4.7)

$$\int_{\partial F_\varrho} |u(x)|\, d\mu_{n-1} \leqq NN_0 \varrho \log \varrho^{-1}$$

for all ϱ with $0 < \varrho \leqq \varrho_0$. Hence we have for $k \leqq n - 2$

$$(4.8) \qquad \lim_{\varrho \to 0} \int_{\partial F_\varrho} |u(x)|\, d\mu_{n-1} = 0 \, .$$

From (4.8) we see that for $k \leqq n - 2$, $u(x)$ satisfies the first condition of c) and condition d). The same is true for $k = n - 1$, since in this case $u(x)$ is continuous in $\mathbf{R}^n$ as a consequence of the supposition of the theorem.

Now the second condition of c) remains. For $k = n - 1$ there is a proof for this condition :· G.A. GARRETT [7]. For $k < n - 1$ one can show that

$$\left|\frac{\partial u(x)}{\partial r}\right| \leqq N\, R^{k-n+1}(x)$$

and thus

$$\int\limits_{\partial F_\varrho} \left|\frac{\partial u}{\partial \nu}\right| d\mu_{n-1} \leqq NN_0 \varrho^{k-n+1} \varrho^{n-1-k} = NN_0$$

for all ϱ with $0 < \varrho \leqq \varrho_0$.

The condition on φ stated in Theorem 2 is for instance fulfilled if φ is absolutely continuous with respect to μ_k and the density of φ is bounded. On the other hand, one can show that under more restrictive assumptions on u(x) than in Theorem 1 the measure φ not only exists but also is absolutely continuous with respect to μ_k. It is then also possible to give concrete expressions for the density of φ. But here we will not go into details.

REFERENCES

[1] S. Dümmel, Einige Eigenschaften von k-dimensionalen λ-Potentialen der einfachen und der doppelten Belegung, Atti Accad. Naz. Lincei, Mem., Cl. Sci. fis. mat. natur., Ser. VIII, 7, 172-201 (1965).

[2] S. Dümmel, Regularitätseigenschaften von k-dimensionalen Potentialen, Bull. Math. Soc. Sci. Math. Phis. R.P.R. 8 (56), 29-38 (1966).

[3] S. Dümmel, Über den Begriff der k-dimensionalen Divergenz, Math. Nachr. 38, 309-321 (1968).

[4] S. Dümmel, Unendlichkeitsstellen von verallgemeinerten Potentialen, Wiss. Zeitschr. Techn. Hochsch. Karl-Marx-Stadt 10, 507-509 (1968).

[5] S. Dümmel, Bemerkungen zur Theorie der k-dimensionalen λ-Potentiale, Wiss. Zeitschr. Techn. Hochsch. Karl-Marx-Stadt 14, 677-680 (1972).

[6] S. Dümmel, k-dimensionale Operatoren der Vektoranalysis, Wiss. Zeitschr. Techn. Hochsch. Karl-Marx-Stadt 15, 43-53 (1973).

[7] G.A. Garrett, Necessary and sufficient conditions for potentials of single and double layers, Amer. J. Math. 58, 95-129 (1936).

[8] W. Rinow, Der Begriff der Ergiebigkeit eines Vektorfeldes und der Gaussche Integralsatz, Bericht über die Mathematikertagung in Berlin vom 14. bis 18. Januar 1953, Berlin 1953, p. 284-289.

[9] L. Schwartz, Théorie des distributions, Tome I, Paris 1957.

Sektion Mathematik, Technische Hochschule, Karl-Marx-Stadt, Reichenhaier Str. 41, GDR

APPLICATION OF ROTHE'S METHOD TO NONLINEAR PARABOLIC BOUNDARY VALUE PROBLEMS

Jozef Kačur
Bratislava (Czechoslovakia)

We shall consider nonlinear parabolic boundary value problems of the form

$$(1)\quad \frac{du(t)}{dt} + A(t)u(t) = f(t); \quad 0 \le t \le T; \quad u(0) = u_o ,$$

where $A(t)$, $t \in \langle 0,T\rangle$ is a system of nonlinear operators.

All the results are obtained on the base of Rothe's method which consists in the following process:

Successively, for $j = 1,\dots,n$, we solve the equations

$$A(t_j)z_j + h^{-1}(z_j - z_{j-1}) = f(t_j) ;$$

where $z_o \equiv u_o$, $t_j = jh$ and $h = Tn^{-1}$.

Then, under some conditions, Rothe's functions z^n

$$[z^n(t) = z_{j-1} + (t-t_{j-1})\, h^{-1}\, (z_j - z_{j-1})$$

for $t_{j-1} \le t \le t_j$, $(j = 1,2,\dots,n)]$ converge for $n \to \infty$ to the solution of the problem (1).

This method was introduced by E. ROTHE in [5] and later has been used by many authors - see e.g. [6], [7], [8], [9], [4], [1] [2], [3] etc.

1. STATIONARY CASE $A(t) \equiv A$

In [1], (part I), it was proved that Rothe's functions converge to the weak solution of the first boundary value problem for the equation

$$\frac{du}{dt} + \sum_{|i| \le k} (-1)^{|i|} D^i a_i(x,Du) = f(x,t) \quad \text{in } \Omega \times \langle 0,T\rangle$$

not only in the usually used Sobolev spaces but also in Orlicz-Sobolev spaces. This result allows to apply the above mentioned method to the differential operators whose coefficients satisfy more general growth conditions.

The abstract form of the results may be formulated as follows - see [2]:
Let H be a real Hilbert space. A nonlinear operator A is defined in $D(A) \subset H$. We suppose

(2) for $u, v \in D(A)$. $(Au-Av,u-v) \ge 0$;

(3) $(A + I)\ (D(A)) = H$;

(4) $f(t) \in C(\langle 0,T\rangle,H)$ with bounded variation in the interval $\langle 0,T\rangle$.

Theorem. *Let us assume* (2) - (4). *If* $u_0 \in D(A)$ *then there exists a unique solution* $u(t)$ *of* (1) *with the following properties:*

$u(t)$ *is Lipschitz continuous from* $\langle 0,T\rangle$ *into* H ,
$u(t) \in D(A)$ *for all* $t \in \langle 0,T\rangle$,
$z^m(t) \to u(t)$ *in* H *uniformly for* $t \in \langle 0,T\rangle$,
$A\,u(t)$ *is weakly continuous in* $t \in \langle 0,T\rangle$.

If u,v *are two solutions corresponding to* f,u_0 *and* g,v_0 *respectively, which are weakly differentiable in* t *for* $t \in \langle 0,T\rangle$, *then*

$$\max_{\langle 0,T\rangle} \|u(t) - v(t)\| \le 2 \int_0^T \|f(t) - g(t)\|\, dt + \|u_0 - v_0\| .$$

If $f(t) \equiv 0$ *then* $\|Au(t)\|$ *is nonincreasing.*

Sketch of the proof: The elements z_j defined above by Rothe's method are uniquely determined because of (2) and (3). Immediately we get

$$(5) \quad h^{-1} \|z_j - z_{j-1}\| \leq \underset{\langle 0,T\rangle}{\mathrm{Var}}\, f + \underset{\langle 0,T\rangle}{\mathrm{Max}} \|f(t)\| + \|Az_0\|$$

and for Rothe's function $z^n(t)$,

$$(6) \quad \|z^n(\tau) - z^n(\mu)\| \leq |\tau - \mu| \left(\underset{\langle 0,T\rangle}{\mathrm{Var}}\, f + \underset{\langle 0,T\rangle}{\mathrm{Max}} \|f(t)\| + \|Az_0\|\right) .$$

Let us define also $x^n(t)$ putting $x^n(0) = u_0$, $x^n(t) = z_j$ for $(j-1)h < t \leq jh$. In the same manner we define $f^n(t)$. It follows from the construction that

$$(7) \quad \int_0^t A(x^n(\tau))\, d\tau + z^n(t) = \int_0^t f^n(\tau)\, d\tau + u_0$$

and that $x^n(t) - z^n(t) \to 0$ uniformly in $\langle 0,T\rangle$.

From (5) - (7) we get easily (d^-/dt is the derivative from the left)

$$(8) \quad \frac{d^-}{dt} \|z^n(t) - z^m(t)\|^2 \to 0 \quad \text{for } n,m \to \infty$$

so $z^n(t) \to u(t)$ uniformly in $\langle 0,T\rangle$. Because of the condition (2), the operator A is maximal monotone, i.e., if for all v in D(A) $(Av - w, v - u) \geq 0$, then $u \in D(A)$ and $Au = w$. It is easy to see that $Ax^n(t) \to Au(t)$, so the maximal monotonicity of A permits to pass in (7) to the limit and we get for every $v \in H$

$$(9) \quad \int_0^t (Au(\tau),v)\, d\tau + (u(t),v) = \int_0^t (f(\tau),v)\, d\tau + (u_0,v) .$$

We can pass to the limit in (6) and since it is

$$\frac{d}{dt} \|u(t) - v(t)\|^2 \leq 2\|f(t) - g(t)\| \, \|u(t) - v(t)\|$$

for the solutions u,v corresponding to f,u_0 and g,v_0 respectively, we get the desired inequality and hence the uniqueness. The rest of the proof is easy.

2. NONSTATIONARY CASE

The results of this section generalize and strengthen the results from [1], (part II). For the details and proof see [3].

Assumptions:

V is a real reflexive Banach space and V' is its dual.

H is a real Hilbert space and $V \cap H$ is a dense set in V and H with the corresponding norms.

(10) $A(t) : V \to V'$ is continuous for all $t \in \langle 0,T \rangle$.

(11) $(A(t)\,u - A(t)\,v,\, u - v) \geq 0$ for u,v V, $t \in \langle 0,T \rangle$.

(12) There exists a nondecreasing function $r(s)$ on the interval $\langle 0, \infty)$ with $\lim_{s \to \infty} r(s) = +\infty$ and such that the inequality $(A(t)\,u,u) \geq \|u\|_V\, r(\|u\|_V)$ holds for all $u \in V$ and $t \in \langle 0,T \rangle$.

(13) Let $t \in \langle 0,T \rangle$ be arbitrary. Suppose
$$A(t)\,u = \operatorname{grad} \Phi(t,u)$$
for $u \in V$, i.e., $A(t)$ are potential operators.

(14) For every $u \in V, t \in (0,T)$ there exist derivatives $A'(t)u$, $A''(t)u$ of $A(t)u$ with respect to the variable t and
$$\|A'(t)u\|_{V'} + \|A''(t)u\|_{V'} \leq C_1 + C_2 r(\|u\|_V)\,.$$

(15) Let $f(t)$ be a Lipschitz function from $\langle 0,T \rangle$ into H.

Theorem? *Let (10) – (15) be fulfilled. If* $u_0 \in V \cap H$ *and* $A(0)\,u_0 \in H$ *then there exists a unique solution of (1) with the following properties:*

$u(t)$ *is Lipschitz continuous from* $\langle 0,T \rangle$ *into* H ,

$u(t) \in L_\infty(\langle 0,T \rangle, V \cap H)$ *and* $u(t)$ *is weakly continuous* in $V \cap H$,

$A(t)\,u(t) \in H$ *for* $t \in \langle 0,T \rangle$ *and* $A(t)\,u(t)$ *is weakly continuous in* H *with respect to the variable* t,

$z^n(t) \to u(t)$ *in* H *uniformly for* $t \in \langle 0,T \rangle$,

$z^m(t) \to u(t)$ *in* $V \cap H$,

$\frac{du(t)}{dt} \in L_\infty(\langle 0,T \rangle, H)$

and

$$\max_{\langle 0,T \rangle} \|z^n(t) - u(t)\|^2 \leq C(u_0,f)\, n^{-1}.$$

If u,v *are two solutions corresponding to* f, u_0 *and* g, v_0 *respectively, then*

$$\max_{\langle 0,T\rangle} \|u(t) - v(t)\| \le \|u_0 - v_0\| + 2 \int_0^T \|f(s) - g(s)\| \, ds \; .$$

REFERENCES

[1] J. Kačur, Method of Rothe and nonlinear parabolic equations of arbitrary order. I and II. Czech. Mat. J., to appear

[2] J. Nečas, Application of Rothe's method to abstract parabolic equations, Czech. Math. Journal (to appear)

[3] J. Kačur, Application of Rothe's method to nonlinear evolution equations. Mat. Časopis Sloven. Akad. Vied, to appear

[4] П.П. Мосолов, Вариационные методы в нестационарных задачах. (Параболический случай) Изв. АН СССР., 34 (1970), 425-457

[5] E. Rothe, Zweidimensionale parabolishe Randwertaufgaben als Grenzfall eindimensionaler Randwertaufgaben. Math. Ann. 102, 1930

[6] О.А. Олейник, Т.Д. Вентцель, Задача Коши и первая краевая задача для квазилинейного уравнения параболического типа. ДАН СССР 97, №4 (1954), 605-608

[7] О.А. Ладыженская, Решение в целом первой краевой задачи для квазилинейных параболических уравнений. ДАН СССР 107, 1956, 636-639

[8] А.М. Ильин, А.С. Калашников, О.А. Олейник, Линейные уравнения второго порядка параболического типа. УМН 17, вып. 3, 1962, 3-146

[9] K. Rektorys, On application of direct variational methods to the solution of parabolic boundary value problems of arbitrary order in the space variables. Czech. Math. J. 21 (96) 1971, 318-339

[10] F.E. Browder, Existence theorems for nonlinear partial differential equations, Global Analysis, Proc. Symp. Pure Math., Vol. 16, Amer. Math. Soc. 1970, 1-60

[11] H. Brezis, Opérateurs Maximaux Monotones, Mathematics Studies, North-Holland, 1973

Prírodovedecká fakulta UK, Mlynská dolina, 816 00 Bratislava 16, Czechoslovakia

POTENTIALS AND REMOVABILITY OF SINGULARITIES

Josef Král
Praha (Czechoslovakia)

We shall deal with results concerning removability of singularities of certain solutions of partial differential equations. Let us first briefly recall what one usually understands by a set of removable singularities. Let $U \subset \mathbf{R}^N$ be an open set and consider a differential operator of the form

$$(1) \quad P(D) = \sum_{\alpha \in M} a_\alpha D^\alpha$$

acting on distributions in U; here M is a finite set of multi-indices $\alpha = [\alpha_1, \ldots, \alpha_N]$ and we write, as usual, $D = D_1^{\alpha_1} \ldots D_N^{\alpha_N}$, where $D_k = -i\partial_k$ and ∂_k denotes differentiation with respect to the k - th variable. For the sake of simplicity we shall always suppose that a_α are infinitely differentiable functions in U or complex constants. Given a class K(U) of distributions in U, then a relatively closed subset $F \subset U$ is termed removable for K(U) with respect to P(D) provided P(D)u = 0 in the whole U whenever $u \in K(U)$ satisfies P(D)u = 0 in $U \setminus F$. Results concerning removability of singularities usually introduce an adequate measure of massiveness of a set and show that sets which are not very massive are removable. As an example we shall mention here a theorem of R.HARVEY and J. POLKING [27] dealing with the class $K(U) = K_\delta(U)$ consisting of functions satisfying locally in U the Hölder condition with exponent δ, $0 < \delta < 1$. It appears that Hausdorff measures serve as a suitable measure of massiveness in this case: If P(D) has order m and $(N-m+\delta)$-dimensional Hausdorff measure of F vanishes, then F is removable for $K_\delta(U)$ with respect to P(D).

This theorem was preceded by results of L. CARLESON [6], [7] and E.P. DOLŽENKO [16], [17] who employed Hausdorff measures in the case when $P(D) = \Delta$ is the Laplace operator. Carleson's result is sharp: It says that a compact subset of U is removable for $K_\delta(U)$ with respect to Δ if and only if its $(N-2+\delta)$-dimensional Hausdorff measure vanishes. It appears that a sharp result of this type holds for the heat conduction operator

$$\Omega = -i\partial_N + \sum_{k=1}^{N-1} \partial_k^2$$

if we consider, instead of $K_\delta(U)$, the class of functions satisfying the anisotropic Hölder condition with exponent $\frac{1}{2}\delta$ in the time variable and exponent δ in the space variables. Reflecting the fact that Ω involves only the first order derivative with respect to the time variable one has also to use a suitable anisotropic modification of the Hausdorff measure resulting from coverings by special parallelepipeds rather than cubes or spheres (compare [32]). This indicates that results of this sort might hold for certain operators of the form (1) if one took into account the shape of the set M of all multiindices which enter P(D) with a non-zero coefficient. Since M is finite it is always possible to choose positive integers $m_1,\dots,m_N$ representing components of a vector denoted by $m = [m_1,\dots,m_N]$ in such a way that M lies below the hyperplane with the equation

$$|x : m| \equiv \sum_{k=1}^{N} \frac{x_k}{m_k} = 1 .$$

We shall always suppose that

$$(2) \qquad \sum_{k=1}^{N} \frac{1}{m_k} > 1$$

which means that the numbers m_k have not been chosen exceedingly large. (Of course, (2) can only be fulfilled if the order of P(D) is small in comparison with the dimension N of the space. For example, in the case of the heat conduction operator Ω the most economic choice of m is $m_N = 1$ and $m_k = 2$ for $k < N$, so that $\sum_{k=1}^{N} \frac{1}{m_k} = \frac{1}{2}(N+1)$ and (2) holds whenever $N > 1$.)

Further we put for $x \in \mathbf{R}^N$

$$|x|_m = \sum_{k=1}^{N} |x_k|^{m_k}$$

and introduce the anisotropic modulus of continuity of a function u in U as follows. If $Q \subset U$ and $r \geq 0$, we put

$$\omega_m^u(r;Q) = \sup\{|u(x)-u(y)| ; x, y \ Q, |x-y|_m \leq r\} .$$

The function

$$r \mapsto \omega_m^u(r;Q)$$

will be termed the modulus of continuity of type m of u on Q.

In what follows we shall always suppose that ω is a non--decreasing continuous function on an interval $\langle 0, \delta_o \rangle$, $\delta_o > 0$, $\omega(0) = 0$. For any open $U \subset \mathbf{R}^N$ we shall denote by U_m^ω the class of all functions u on U satisfying

$$\omega_m^u(r;Q) = \mathcal{O}(\omega(r)) \quad \text{as } r \to 0+$$

for every compact $Q \subset U$.

Next we are going to define the anisotropic Hausdorff measures of type m. By a lump of type m we shall mean any set of the form

$$K = \bigtimes_{j=1}^{N} I_j ,$$

where each I_j is a one-dimensional interval of length r^{1/m_j}; the corresponding number $r > 0$ will be termed the norm of K and will be denoted by $|K|_m$. (It should be noted here that, in connection with removability of singularities, W. LITTMAN [35] employed lumps of this type for constructing measures analogous to the Minkowskian content.) If φ is a continuous non-decreasing function (= measure function) on an interval $\langle 0, \varepsilon_o \rangle$, $\varepsilon_o > 0$, $\varphi(0) = 0$, we put for $A \subset \mathbf{R}^N$ and $\varepsilon \in (0, \varepsilon_o)$

$$H^{\varepsilon}_{m,\varphi}(A) = \inf \sum_{n} \varphi(|K_n|_m) ,$$

where the infimum is taken over all sequences of lumps K_n of type m such that

$$\bigcup_{n} K_n \supset A , \quad |K_n|_m \le \varepsilon \ \forall n .$$

Finally we define the Hausdorff measure of type m corresponding to the measure function φ by

$$H_{m,\varphi}(A) = \lim_{\varepsilon \to 0+} H^{\varepsilon}_{m,\varphi}(A) ,$$

as it is usual in the theory of Hausdorff measures (cf. [47]). If, in particular, $\varphi(t) = t^p$, then $H_{m,\varphi}$ will be denoted by H^p_m and will be termed the p-dimensional measure of type m.

Now we are in a position to formulate the following

1. Lemma. *Consider an operator of the form (1) with infinitely differentiable coefficients* a_α *in* U, *suppose that* $|\alpha : m| \le 1$ *whenever* $\alpha \in M$ *and let*

$$(3) \qquad \beta = \perp + \sum_{k=1}^{N} \frac{1}{m_k} > 0 .$$

Let

$$(4) \qquad \omega(2r) = \mathcal{O}(\omega(r)) \quad \text{as } r \to 0+$$

and put

$$(5) \qquad \varphi(r) = r^{\beta}\, \omega(r) , \quad r \in \langle 0, \delta_0 \rangle .$$

Let $F \subset U$ *be relatively closed and nowhere dense in* U. *If* $u \in U^{\omega}_{m}$ *satisfies* $P(D)u = 0$ *in* $U \setminus F$, *then for every compact* $Q \subset U$ *there is a constant* k_Q *such that, for every infinitely differentiable function* ψ *with support contained in* Q, $|\langle P(D)u, \psi\rangle| \le k_Q H_{m,\varphi}(F \cap Q) \max |\psi(Q)|$.

Proof of this lemma is obtained by slight modifications of reasonings occurring in [34], Section 14 and will be omitted here. We shall only mention that it rests on a generalization of the fundamental Lemma 3.1 of R. HARVEY and J. POLKING [27] (cf. Lemma 13 in [34]). The above Lemma implies the following

2. Theorem. *Let* P(D), U, F *have the same meaning as in Lemma 1, assume* (3), (4) *and define* φ *by* (5). *If* F *has locally finite* $H_{m,\varphi}$*-measure then, for every* $u \in U_m^\omega$ *satisfying* $P(D)u = 0$ *in* $U \setminus F$, *there is a locally bounded Baire function* g_u *in* U *vanishing on* $U \setminus F$ *such that*

$$P(D)u = g_u \, H_{m,\varphi} \text{ in } U$$

(*which means that, for every infinitely differentiable function* ψ *with compact support in* U, $\langle P(D)u, \psi\rangle = \int_U \psi g_u \, d\, H_{m,\varphi}$). *In particular, if* $H_{m,\varphi}(F) = 0$, *then* F *is removable for* U_m^ω *with respect to* P(D).

The theorem presents a sufficient condition for removability of a singular set. In order to obtain necessary conditions for removability we are first going to describe certain properties of potentials derived from general kernels including those resulting from fundamental solutions of semielliptic equations with constant coefficients. In what follows we shall always assume that G(x,y) is a complex valued Baire function of the variables $x, y \in \mathbf{R}^N$ which is bounded on every set of the form

$$\{[x, y];\ x, y \in \mathbf{R}^N,\ \varepsilon < |x-y| < \varepsilon^{-1}\},\quad 0 < \varepsilon < 1.$$

A simple calculation (compare Lemma 2 in [34]) yields the following

3. Lemma. *Assume* (3) *and suppose that*

$$(6) \quad G(x,y) = \mathcal{O}(|x-y|_m^{-\beta}) \quad \textit{as } |x-y| \to 0;$$

Let φ *be a continuous non-decreasing function on* $\langle 0, \delta_0\rangle$, $\delta_0 > 0$, $\varphi(0) = 0$, *and suppose that*

$$\varphi(2r) = \mathcal{O}(\varphi(r)) \quad \textit{as } r \to 0+ .$$

Let μ be a compactly supported Borel measure in $\mathbf{R}^N$ *satisfying the estimate*

$$(7) \quad \mu(K) \le \varphi(|K|_m)$$

for every lump K *of type* m *with* $|K|_m \le \delta_0$. *Then the potential*

$$G\mu(x) = \int_{\mathbf{R}^N} G(x,y)\, d\mu(y)$$

is defined for all $x \in \mathbf{R}^N$ *and, for any fixed* $s > 0$,

$$\int_{|x-y|_m < sr} |G(x,y)|\, d\mu(y) = \mathcal{O}(\varphi(r) r^{-\beta} + \int_0^r \varphi(t)\, t^{-\beta-1}\, dt) \text{ as } r \to 0+.$$

4. Lemma. *Let* $G(x,y)$ *be continuously differentiable with respect to* x *outside the diagonal* $\{[x,x];\ x \in \mathbf{R}^N\}$ *and suppose that, for* $j = 1, \ldots, N$,

$$\frac{dG(x,y)}{dx_j} = \mathcal{O}\,(|x-y|_m^{-\beta-1/m_j}) \quad \text{as } |x-y| \to 0+ ,$$

where β *fulfils* (3). *Further suppose that* φ, μ *satisfy the assumptions of Lemma 3. Then* (6) *holds so that, by Lemma 3, the potential* $u(x) = G\mu(x)$ *is everywhere defined. Besides that, its modulus of continuity of type* m *admits, on every compact* $Q \subset \mathbf{R}^N$, *the estimate*

$$\omega_m^u(r,Q) = \mathcal{O}\Big(\int_0^r \varphi(t)\, t^{-\beta-1}\, dt + \sum_{j=1}^N r^{1/m_j} \,.$$

$$\int_r^{\delta_0} \varphi(t)\, t^{-\beta-1-1/m_j}\, dt\Big) \quad as \quad r \to 0+ .$$

Proof of this Lemma follows from reasonings similar to those employed in Section 4 in [34].

In connection with the applicability of Lemma 4 it is useful to know which compact sets may serve as supports of non-trivial measures μ fulfilling (7) for all sufficiently small lumps K of type m. Such compacts are characterized by the following slight modification of a classical result of O. FROSTMAN [23].

5. Lemma. *If $Q \subset \mathbf{R}^N$ is compact and φ has the same meaning as in Lemma 3, then $H_{m,\varphi}(Q) > 0$ is a necessary and sufficient condition for the existence of a non-trivial Borel measure μ with support contained in Q fulfilling (7) for all sufficiently small lumps K of type m.*

Lemmas 4 and 5 may be combined to yield conditions on the set $F \subset U$ guaranteeing the existence of a non-trivial measure μ supported by F such that the corresponding potential $G\mu$ belongs to U_m^ω.

6. Proposition. *Let G fulfil the assumptions described in Lemma 4. Let ω satisfy (4), suppose that there is an $\varepsilon > 0$ such that all the functions,*

$$t \mapsto \omega(t)\, t^{\varepsilon - 1/m_j} \quad (j=1,\dots,N)$$

are non-increasing on $(0,\varepsilon_1)$ for a suitable $\varepsilon_1 \in (0,\varepsilon_0)$ and, besides that, ω is continuously differentiable on $(0,\varepsilon_1)$, the function

$$t \mapsto t^{\beta+1}\, \omega'(t)$$

is non-decreasing on $(0,\varepsilon_1)$ and tends to zero as $t \to 0+$. Put $\varphi(0) = 0$, $\varphi(t) = t^{\beta+1}\omega'(t)$ for $t \in (0,\varepsilon_1)$. If $U \subset \mathbf{R}^N$ is open and F is a relatively closed subset of U with $H_{m,\varphi}(F) > 0$, then there is a non-trivial Borel measure μ with compact support contained in F such that $G\mu \in U_m^\omega$.

Remarks. If G is the Riesz kernel then sharp conditions on μ guaranteeing that the potential $G\mu$ satisfies the Hölder condition have been obtained by H. WALLIN [61].

Investigations of E.P. DOLŽENKO [17] indicate that, in the above proposition, the assumption concerning differentiability of ω is actually not restrictive.

If the kernel G derived from the fundamental solution corresponding to a differential operator P(D) fulfils the estimates required in Proposition 6, then this proposition implies that, under appropriate assumptions on ω, $H_{m,\varphi}(F) = 0$ (with $\varphi(t) =$

$= t^{\beta+1}\ \omega'(t))$ is necessary for F to be removable for U_m with respect to P(D).

Let us recall that a differential operator with constant coefficients of the form

$$(8) \quad P(D) = \sum_{|\alpha:m|\leq 1} a_\alpha D^\alpha$$

is termed semielliptic if the leading polynomial

$$P_m(\xi) = \sum_{|\alpha:m|=1} a_\alpha \xi^\alpha$$

(where, for $\xi \in \mathbf{R}^n$, $\xi^\alpha = \xi_1^{\alpha_1} \dots \xi_N^{\alpha_N}$) corresponding to the generalized principal part of P(D) has no non-trivial zero point in $\mathbf{R}^N$, i.e.,

$$(\xi \in \mathbf{R}^N,\ P_m(\xi) = 0) \Longrightarrow \xi = 0 .$$

It is well known that in this case the vector m is uniquely determined, m_k being just the degree of P(D) in the variable D_k (cf. [29], [58]). For semielliptic operators we get the following corollary of Proposition 6.

7. Theorem. *Let* P(D) *be a semielliptic operator of the form* (8) *with constant (complex) coefficients. Assume* (3) *and suppose that* ω *satisfies the conditions described in Proposition 6. As in Proposition 6, let* $\varphi(t) = t^{\beta+1}\ \omega'(t)$ *for* $t \in (0,\varepsilon_1)$, $\varphi(0) = 0$. *If* $U \subset \mathbf{R}^N$ *is open, then a relatively closed set* $F \subset U$ *is removable for* U_m^ω *with respect to* P(D) *only if* $H_{m,\varphi}(F) = 0$.

P r o o f. Let E be the fundamental solution corresponding to P(D). Since P(D) is hypoelliptic, E is infinitely differentiable on $\mathbf{R}^N$ {0} (cf. [29], [58]). If $G(x,y) = E(x,y)$ for $x \neq y$, then G fulfils the conditions described in Lemma 4, as follows from results of V.V. GRUŠIN [26]. For definiteness, let $G(x,x) = 0$ whenever $x \in \mathbf{R}^N$. If F is a relatively closed subset of U with $H_{m,\varphi}(F) > 0$ then, by Proposition 6, there is a non-trivial Borel measure μ compactly supported by F such that the corresponding potential $u = G\mu$ belongs to U_m^ω. Now it is sufficient to realize that $u = E * \mu$ (= the convolution of E and μ) satisfies $P(D)u = \mu$ in the sense of distribution theory.

8. Remark. Let us keep the assumptions on P(D) and ω introduced in the preceding section. Theorem 7 shows that a necessary condition for removability of F for U_m^ω reads

$$(9) \quad H_{m,\varphi_n}(F) = 0 \ , \quad \text{where} \quad \varphi_n(t) = t^{\beta+1}\,\omega'(t) \ ,$$

while Theorem 2 gives the following sufficient condition for F to be removable for U_m^ω:

$$(10) \quad H_{m,\varphi_s}(F) = 0 \ , \quad \text{where} \quad \varphi_s(t) = t^{\beta}\,\omega(t) \ .$$

There is a broad class of functions ω for which the conditions (9), (10) are equivalent to each other. Theorems 2 and 7 can then be combined to yield a necessary and sufficient condition for removability of a set for U_m^ω. E.g., this happens when $\omega(t) = t^\gamma \log^\varrho \frac{1}{t}$ with $\gamma \in (0,1)$ and $\varrho \in \mathbf{R}^1$.

Many results dealing with removable singularities and various measures of exceptional sets may be found in the books and papers listed below.

REFERENCES

[1] D.R. Adams, N.G. Meyers, Bessel potentials. Inclusion relations among classes of exceptional sets, Bull. Amer. Math. Soc. 77 (1971), 968-970

[2] R.A. Adams, Properties of equivalent capacities, Canad. Math. Bull. 14 (1971), 5-11

[3] G. Anger, Funktionalanalytische Betrachtungen bei Differentialgleichungen unter Verwendung von Methoden der Potentialtheorie I, Akademie - Verlag, Berlin 1967

[4] D.G. Aronson, Removable singularities for linear parabolic equations, Arch. Rational Mech. Anal. 17 (1964), 79-84

[5] S. Bochner, Weak solutions of linear partial differential equations, J.Math. Pures Appl. 35 (1956), 193-202

[6] L. Carleson, Removable singularities of continuous harmonic functions in R^m, Math. Scand. 12 (1963), 15-18

[7] L. Carleson, Selected problems on exceptional sets, Van Nostrand 1967

[8] V.P. Chavin, S.Ja. Chavinson, Někotoryje ocenki analitičeskoj jemkosti, Doklady AN SSSR 138 (1961), 789-792

[9] V.P. Chavin, V.G.Mazja: see V.G. Mazja, V.P. Chavin

[10] V.P. Chavin, R.E. Val'skij: see R.E. Val'skij, V.P. Chavin

[11] S.Ja. Chavinson, O stiranii osobennostěj, Litovskij matem. sbornik 3 (1963), 271-287

[12] S.Ja. Chavinson, V.P. Chavin: see V.P. Chavin, S.Ja. Chavinson

[13] J. Diederich, Removable sets for pointwise solutions of elliptic partial differential equations, Trans. Amer. Math. Soc. 165 (1972), 333-352

[14] J.R. Diederich, V.L. Shapiro, Removable sets for pointwise C^{α} solutions of elliptic partial differential equations, Journal of differential equations 11 (1972), 562-581

[15] E.P. Dolženko, O stiranii osobennostěj analitičeskich funkcij, Usp. mat. nauk 18 (1963), 135-143

[16] E.P. Dolženko, O predstavlenii něpreryvnych garmoničeskich funkcij v vidě potěncialov, Izvestija Akad. Nauk SSSR 28 (1964), 1113-1130

[17] E.P. Dolženko, Ob osobych točkach něpreryvnych garmoničeskich funkcij, ibid. 1251-1270

[18] M. Dont, Removable singularities for solutions of an equation, Acta Univ. Carolinae, to appear

[19] R.E. Dyer, D.E. Edmunds, Removable singularities of solutions of the Navier-Stokes equations, J. London Math. Soc. 2 (1970), 535-538

[20] D.E. Edmunds, R.E. Dyer: see R.E. Dyer, D.E. Edmunds

[21] D.E. Edmunds, L.A. Peletier, Removable singularities of solutions of quasilinear parabolic equations, J. London Math. Soc. (2) 2 (1970), 273-283

[22] D.E. Edmunds, L.A. Peletier, Removable singularities of solutions of degenerate elliptic equations, Bolletino U.M.I. (4) 5 (1972), 345-356

[23] O. Frostman, Potentiel d'équilibre et capacité des ensembles, Meddel. Lunds. Univ. Mat. Sem. 3 (1935)

[24] B. Fuglede, On the axiomatic theory of thin sets in potential theory, Vorträge d. 3. Tagung über Probleme u. Methoden d. Math. Physik, Karl-Marx--Stadt 1966

[25] J. Garnett, Positive length but zero analytic capacity, Proc. Amer. Math. Soc. 24 (1970), 696-699

[26] V.V. Grušin, Svjaz' meždu lokal'nymi i global'nymi svojstvami rešenij gipoelliptičeskich uravněnij s postojannymi koeficientami, Matem. sbornik 66 (108) (1965), 525-550

[27] R. Harvey, J. Polking, Removable singularities of solutions of linear partial differential equations, Acta Mathematica (Uppsala) 125 (1970), 39-56

[28] R. Harvey, J. Polking, A notion of capacity which characterizes removable singularities, Trans. Amer. Math. Soc. 169 (1972), 183-195

[29] L. Hörmander, Linear partial differential operators, Springer-Verlag 1963

[30] R. Infantino, Sulle singolarità eliminabili per le soluzioni deboli delle equazioni lineari ellittiche del quarto ordine, Ricerche Mat. 18 (1969), 102-119

[31] R. Infantino, Un teorema di singolarità eliminabili per le soluzioni deboli delle equazioni lineari ellittiche, Atti Accad. Naz. Lincei, Rendiconti Cl. Sc. fis., mat. e natur. L II (1972), 319-326

[32] J. Král, Hölder-continuous heat potentials, Atti Accad. Naz. Lincei, Rendiconti Cl. Sc. fis., mat. e natur. L I (1971), 17-19

[33] J. Král, Regularity of potentials and removability of singularities of solutions of partial differential equations, Proc. Conference Equadiff 3 (held in Brno, August 28-September 1, 1972), 179-185

[34] J. Král, Removable singularities of solutions of semielliptic equations, Rendiconti di Matematica, to appear

[35] W. Littman, Polar sets and removable singularities of partial differential equations, Ark. Mat. 7 (1967), 1-9

[36] W. Littman, A connection between α-capacity and m-p polarity, Bull. Amer. Math. Soc. 73 (1967), 862-866

[37] V.G. Maz'ja, V.P. Chavin, Priloženija (p,l) - jemkosti k něskol'kim zadačam teorii isklučitel'nych množestv, Matem. sbornik 90 (132) (1973), 558--591.

[38] N.G. Meyers, A theory of capacities for potentials of functions in Lebesgue classes, Math. Scand. 26 (1970), 255-292

[39] N.G. Meyers, D.R. Adams: see D.R. Adams, N.G. Meyers

[40] C. Miranda, Partial differential equations of elliptic type, Springer-Verlag, 1970

[41] M. Ohtsuka, On various definitions of capacity and related notions, Nagoya Math. J. 30 (1967), 121-127

[42] L.A. Peletier, D.E. Edmunds: see D.E. Edmunds, L.A. Peletier

[43] Ju.A. Peškičev, Někotoryje svojstva α-jemkosti množestv, Sibirskij mat. žurnal XII (1971), 613-622

[44] B. Pini, Sui punti singolari delle soluzioni delle equazioni paraboliche lineari, Ann. dell' Univ. di Ferrara, Sez. VII, vol. 2 (1953), 2-12

[45] J. Polking, R. Harvey: see R. Harvey, J. Polking

[46] Ju.G. Rešetnjak, O množestve osobych toček rešenij někotorych nělinějnych uravněnij elliptičeskogo tipa, Sibirsk. Mat. Ž. 9 (1968), 354-367

[47] C.A. Rogers, Hausdorff measures, Cambridge Univ. Press 1970

[48] B.W. Schulze, Potentiale bei der Wellengleichung im R^2 und Charakterisierung der Mengen der Kapazität Null, Elliptische Differentialgleichungen (Tagungsberichte, herausgeben von G. Anger, Akademie-Verlag Berlin, 1971) Bd. I, 137-157

[49] B.W. Schulze, Nullmengensysteme in der Potentialtheorie, Math. Nachrichten 49 (1971), 293-309

[50] J. Serrin, Removable singularities of solutions of elliptic equations I, II, Archive Rational Mech. Anal. 17 (1964), 67-78, ibid. 20 (1965), 163-169

[51] J. Serrin, Local behavior of solutions of quasi-linear equations, Acta Math. 111 (1964), 247-302

[52] J. Serrin, Singularities of solutions of nonlinear equations, Proc. Symposia in Appl. Math. vol. XVII, Amer. Math. Soc. 1965, 68-88

[53] V.L. Shapiro, Removable sets for pointwise solutions of the generalized Cauchy-Riemann equations, Annals of Math. 92 (1970), 82-101

[54] V.L. Shapiro, Removable sets for pointwise subharmonic functions, Trans. Amer. Math. Soc. 159 (1971), 369-380

[55] V.L. Shapiro, Capacity and the non-linear Navier-Stokes equations, Siam. J. Math. Anal. 4 (1973), 329-343

[56] V.L. Shapiro, J.R. Diederich: see J.R. Diederich, V.L. Shapiro

[57] S.J. Taylor, On the connection between Hausdorff Measures and generalized capacity, Proc. Cambridge Philos. Society (Math. and Phys. Sc.) 57 (1961), 524-531

[58] F. Trèves, Lectures on partial differential equations with constant coefficients, Notas de Matematica N° 7, Rio de Janeiro 1961

[59] R.E. Val'skij, V.P. Chavin, Stiranije osobennostěj analitičeskich funkcij i peremeščenija mass, Sibirskij mat. žurnal VII (1966), 55-60

[60] A.G. Vituškin, Primer množestva položitěl'noj dliny no nulevoj analitičeskoj jemkosti, Doklady AN SSSR 127 (1959), 246-249

[61] H. Wallin, Existence and properties of Riesz potentials satisfying Lipschitz condition, Math. Scand. 19 (1966), 151-160

[62] H. Wallin, Riesz potentials, k, p - capacity and p-module, to appear

[63] G. Wildenhain, Über einen Kapazitätsbegriff bei der Differentialgleichung $\Delta^m u = 0$, Math. Zeitschr. 96 (1967), 125-142

[64] W.P. Ziemer, Extremal length and p-capacity, Michigan Math. J. 16 (1969), 43-51

Matematický ústav ČSAV, Žitná 25, 115 67 Praha 1, Czechoslovakia

THEOREM OF FRÉCHET AND ASYMPTOTICALLY ALMOST PERIODIC SOLUTIONS OF SOME NONLINEAR EQUATIONS OF HYPERBOLIC TYPE

Vladimír Lovicar
Praha (Czechoslovakia)

The aim of this lecture is to prove the existence of asymptotically almost periodic solutions of a second order equation with nonlinear dissipative term by an easy method based on a theorem of Fréchet. For the sake of simplicity, very restrictive assumptions on the coefficients of the equation are introduced. A more general situation is considered in paper [5], where the reader can find also the proofs of Lemmas 1-4, used in this lecture. The existence of asymptotically almost periodic solutions gives us the possibility to prove the existence of almost periodic solutions. For an other method for obtaining almost periodic solutions of equations of the same type as in this lecture see [2]. As in the method mentioned above, the proof of existence of bounded solutions plays here the crucial role. It is very well known that their existence may be derived from local existence of solutions and from appropriate apriori estimates. The local existence of solutions will be supposed here, since this problem is treated in many other papers.

1. NOTATION, AUXILIARY LEMMAS

For any Banach space X (over a field K where K is the set $\mathbf{R}$ of reals or the set $\mathbf{C}$ of complex numbers) we shall denote by $|\cdot|_X$ the norm on X, by X^d the dual space to X and by $\langle .,. \rangle_X$ the duality between X and X^d; for any $a \in K$ and $y \in X^d$ the element $ay \in X^d$ is given by $\langle x, ay \rangle_X = \bar{a} \langle x, y \rangle_X$. If the space X is a Hilbert space (i.e., if the norm on X is given by a scalar product) then we may identify the space X^d with X. For any interval $I \subset \mathbf{R}$ we shall denote by C(I,X) the set of all continuous functions on I into X and by $C_s(I,X)$ the Banach space of bounded functions from C(I,X), equipped with the supremum norm denoted by $|\cdot|_{\infty,X}$. The spaces $L_p(I,X)$ or $L_{p,loc}(I,X)$ are defined as usual.

In all paper we shall suppose that Banach spaces B, H are given (over the same field of scalars) such that $B \subset H$, the embeding being dense and continuous. Further, the space H is supposed to be a Hilbert space and hence after identification of H^d with H we have $H \subset B^d$, the embeding being also dense and continuous. The scalar product on H is denoted by $(.,.)_H$. We shall denote by E the space B x H, equipped with the norm $|[x,y]|_E^2 = |x|_B^2 + |y|_H^2$.

Let A be a linear bounded operator on B into B^d such that $\langle x, Ax \rangle_B = |x|_B^2$ (hence B is in fact isomorphic to a Hilbert space). We shall denote by $\hat{A}$ the operator on C(I,B) into $C(I,B^d)$ defined by $\hat{A}x(t) = A(x(t))$ ($x \in C(I,B)$, $t \in I$). Further, let f,g be functions on R x B x H into H and $\varepsilon \geq 0$. We shall consider the equation

$$(1) \quad x''(t) + Ax(t) = f(t,x(t),x'(t)) + \varepsilon g(t,x(t),x'(t)) .$$

A solution of the equation (1) on an interval $I \subset \mathbf{R}$ is a couple $u = [x,y]$ of functions on I into B, H respectively (i.e. a function u on I into E) for which

a) $x \in C(I,B)$, $y \in C(I,H)$;

b) $x' \in C(I,H)$, $x' = y$;

c) $y' = x'' \in L_{1,loc}(I,B^d)$;

d) the function $t \mapsto h(t) = f(t,x(t),y(t)) + \varepsilon g(t,x(t),y(t))$ belongs to $L_{1,loc}(I,H)$;

e) $x'' + \hat{A}x = h$ (in the sense of the space $L_{1,loc}(I,B^d)$) .

We shall suppose that the solutions of (1) locally exist uniformly in the following sense:

(2) To any $r \geq 0$ there exist $\varepsilon_o(r) > 0$ and $a(r) > 0$ with the following property: if $u_o \in E$, $|u_o|_E - r$, $t \in \mathbf{R}$ and $\varepsilon \in \langle 0, \varepsilon_o(r)\rangle$, then there is a solution u_ε of (1) on the interval $\langle t, t+a(r)\rangle$ with $u_\varepsilon(t) = u_o$.

Lemma 1. *Let* u_1, u_2 *be solutions of (1) on the intervals* I_1, I_2 *respectively (and with the same* ε, *of course). Further, let* $I_1 \cap I_2 \neq \emptyset$ *and let* $u_1 = u_2$ *on* $I_1 \cap I_2$. *Then the function* u *defined by*

$$u(t) = \begin{cases} u_1(t) & (t \in I_1) \\ u_2(t) & (t \in I_2 \setminus I_1) \end{cases}$$

is a solution of (1) on the interval $I = I_1 \cup I_2$.

This Lemma together with the assumption (2) gives us the possibility to prove the existence of global solutions (on large intervals) of the equation (1) if we have some appropriate apriori estimates for solutions of (1). Namely, if we are able to show that to any $r \geq 0$ there is an $\varepsilon_1(r) > 0$ and $r_1 \geq 0$ such that for any solution u_ε of (1) on some interval $\langle I, T\rangle$ with $|u_\varepsilon(0)|_E \leq r$ and $\varepsilon \in \langle 0, \varepsilon_1(r)\rangle$ it holds $|u_\varepsilon(t)|_E \leq r_1$ for $t \in \langle 0, T\rangle$, then we may conclude that to any $|u_o|_E \leq r$ and for sufficiently small $\varepsilon \geq 0$ there exists a solution u_ε of (1) on $\mathbf{R}^+ = \langle 0, +\infty)$ with $u_\varepsilon(0) = u_o$ and $|u_\varepsilon(t)|_E \leq r_1$ for $t \in \mathbf{R}^+$. In the sequel, we shall use this known fact without further notice.

All necessary estimates for solutions of (1) will be obtained by means of relation (3) of the "energy equality" type from the following easy Lemma:

Lemma 2. *Let* $I \subset \mathbf{R}$, $h \in L_{1,loc}(I,H)$ *and let* $u = [x,y]$ *be a solution of the equation* $x''(t) + Ax(t) = h(t)$ *on the interval* I. *Then for any* $a, t \in I$ *and for any real* c *it holds*

$$|u(t)|_E^2 + 2c\,\mathrm{Re}(x(t),y(t))_H = e^{-2c(t-a)}\,(|u(a)|_E^2 +$$

$$(3) \qquad + 2c\,\mathrm{Re}(x(a),y(a))_H) + 2\int_a^t e^{-2c(t-s)}\,\mathrm{Re}(h(s) +$$

$$+ 2\,cy(s), y(s) + cx(s))_H\, ds\ .$$

2. BOUNDED SOLUTIONS OF THE EQUATION (1)

In this section we shall prove that under the assumption (2) and some other additional assumptions on the functions f,g (the assumptions (4) below), solutions of the equation (1) exist and are bounded on $\mathbf{R}^+$. To this aim we shall show that there exists a continuous function F on $\mathbf{R}^+ \times \mathbf{R}^+ \times \mathbf{R}^+$ into $\mathbf{R}$ with the following properties:

a) If u_ε is a solution of (1) on an interval $\langle 0,T\rangle$ and $v_\varepsilon(t) = \sup(|u_\varepsilon(s)|_E;\ s \in \langle 0,t\rangle)$ for $t \in \langle 0,T\rangle$, then

$$F(v_\varepsilon(t), |u_\varepsilon(0)|_E, \varepsilon) \le 0 \quad (t \in \langle 0,T\rangle) ;$$

b) to any $r \ge 0$ there exist $\varepsilon_1 > 0$ and $r_1 > 0$ such that for all $s \in \langle 0,r\rangle$ and $\varepsilon \in \langle 0,\varepsilon_1\rangle$ it holds

$$F(r_1, s, \varepsilon) > 0 .$$

It follows immediately from the above that if u_ε is a solution of (1) on an interval $\langle 0,T\rangle$ with $|u_\varepsilon(0)|_E \le r$ and $\varepsilon \in \langle 0,\varepsilon_1\rangle$ then $|u_\varepsilon(t)|_E \le r_1$ for $t \in \langle 0,T\rangle$, which implies the existence of bounded solutions of (1) on $\mathbf{R}^+$ as it was mentioned in Section 1.

We shall suppose that the functions f,g satisfy the following assumptions:

(4)

a) There exist $k_1 > 0$ and $k_2 \ge 0$ such that

$$\mathrm{Re}(f(t,x,y),y)_H \le -k_1|y|_H^2 + k_2$$

for $t \in \mathbf{R}$, $x \in B$ and $y \in H$;

b) $|f(t,x,y)|_H \le k_3|y|_H + k_4$ for some $k_3, k_4 \ge 0$ and for $t \in \mathbf{R}$, $x \in B$ and $y \in H$;

c) there exists a continuous nondecreasing function m_g on $\mathbf{R}^+$ into $\mathbf{R}^+$ such that

$$|g(t,x,y)|_H \le m_g(|[x,y]|_E)$$

for $t \in \mathbf{R}$ and $[x,y] \in E$.

Let now $u = [x,y]$ be a solution of (1) on the interval $\langle 0,T\rangle$. By Lemma 2 it holds for any real c and for $t \in \langle 0,T\rangle$:

$$|u(t)|_E^2 + 2c\,\mathrm{Re}(x(t),y(t))_H = e^{-2ct}\,(|u(0)|_E^2 +$$

$$+ 2c\,\mathrm{Re}(x(0),y(0))_H) + 2\int_0^t e^{-2c(t-s)}\,h(s)\,ds\ ,$$

where $h(s) = \mathrm{Re}(f(s,x(s),y(s)) + \varepsilon g(s,x(s),y(s)) + 2\,cy(s),y(s) + cx(s))_H$. We obtain from assumptions (4) that for nonnegative c,

$$h(s) \le (2c-k_1)|y(s)|_H^2 + c(k_3+2c)|x(s)|_H\,|y(s)|_H + ck_4|x(s)|_H +$$

$$+ k_2 + \varepsilon m_g(|u(s)|_E)\,(|y(s)|_H + c|x(s)|_H)\ .$$

Let $k > 0$ be such that $|x|_H \le k|x|_B$ for $x \in B$. Further, let $c > 0$ be fixed and such that $|2c\mathrm{Re}(x,y)_H| \le 2^{-1}\,|[x,y]|_E^2$ for $[x,y] \in E$, $2c < k_1$ and $ck^2(k_3+2c)^2(2(k_1-2c))^{-1} < 1$ (in other words, c sufficiently small). Since $(2c-k_1)|y(s)|_H^2 + c(k_3+2c)|x(s)|_H\,|y(s)|_H \le c^2(k_3+2c)^2\,(4(k_1-2c))^{-1}\,|x(x)|_H^2$, the above relations yield easily the estimate

$$|u(t)|_E^2 \le 3|u(0)|_E^2 + 4\int_0^t e^{-2c(t-s)}\,(c^2k^2(k_3+2c)^2$$

$$(4(k_1-2c))^{-1}|u(s)|_E^2 + ckk_4|u(s)|_E + k_2 +$$

$$+ \varepsilon(1+ck)\,m_g(|u(s)|_E)\,|u(s)|_E)\,ds\ .$$

Since $\int_0^t e^{-2c(t-s)}\,ds \le (2c)^{-1}$, one can easily obtain from the above

$$K_1\,v(t)^2 - K_2\,v(t) - K_3 - \varepsilon K_4\,m_g(v(t))\,v(t) - 3|u(0)|_E \le 0\ ,$$

where $v(t) = \sup\,(|u(s)|_E;\ s \in \langle 0,t\rangle)$, $K_1 = 1 - ck^2(k_3+2c)^2(2(k_1-2c))^{-1} > 0$, $K_2 = 2kk_4$, $K_3 = 2c^{-1}k_2$ and $K_4 = 2(c^{-1} + k)$.

Let F denote the function defined on $\mathbf{R}^+ \times \mathbf{R}^+ \times \mathbf{R}^+$ by

$$F(a,b,\varepsilon) = K_1a^2 - K_2a - K_3 - \varepsilon K_4\,m_g(a)a - 3\,b\ .$$

It was stated above that the function F has the property a) from the beginning of this section. It is clear that F has also the property b). This implies the following

Theorem 1. *Let the equation* (1) *be given and let the assumptions* (2) *and* (4) *be satisfied. Then to any* $r \geq 0$ *there exist* $\varepsilon_1(r) > 0$ *and* $r_1 > 0$ *such that*

a) *to any* $u_0 \in E$, $|u_0|_E \leq r$ *and* $\varepsilon \in \langle 0, \varepsilon_1(r)\rangle$ *there is a solution* u_ε *of the equation* (1) *on the interval* $\mathbf{R}^+$ *with* $u_\varepsilon(0) = u_0$;

b) *if* u_ε *is a solution of the equation* (1) *on* $\mathbf{R}^+$ *with* $|u_\varepsilon(0)|_E \leq r$ *and* $\varepsilon \in \langle 0, \varepsilon_1(r)\rangle$, *then* $u_\varepsilon \in C_s(\mathbf{R}^+, E)$ *and* $|u_\varepsilon|_{\infty, E} \leq r_1$.

3. ASYMPTOTICALLY ALMOST PERIODIC SOLUTIONS OF THE EQUATION (1)

Let p_t ($t \in \mathbf{R}$) denote the operator of translation on $\mathbf{R}$, i.e., $p_t x(s) = x(t+s)$ ($s \in \mathbf{R}$) for any function x on $\mathbf{R}$. Further, let p_t^+ ($t \in \mathbf{R}^+$) denote the operator of left translation on $\mathbf{R}^+$, i.e., $p_t^+ x(s) = x(t+s)$ ($s \in \mathbf{R}^+$) for any function x on $\mathbf{R}^+$.

Let X be a Banach space. A function $x \in C_s(\mathbf{R}, X)$ is called almost periodic iff the set $(p_t x;\ t \in \mathbf{R})$ is totally bounded in $C_s(\mathbf{R}, X)$. The set of almost periodic functions from $C_s(\mathbf{R}, X)$ will be denoted by AP($\mathbf{R}$, X). A function $x \in C_s(\mathbf{R}^+, X)$ is called asymptotically almost periodic iff there exists $x_1 \in AP(\mathbf{R}, X)$ such that $\lim_{t \to \infty} |x(t) - x_1(t)|_X = 0$. The set of asymptotically almost periodic functions from $C_s(\mathbf{R}^+, X)$ will be denoted by AAP ($\mathbf{R}^+$, X). In this section we shall use a theorem of Fréchet (see [1]):

Theorem (Fréchet). *Let* X *be a Banach space and let* $x \in C_s(\mathbf{R}^+, X)$. *Then the following conditions are equivalent:*

1. $x \in AAP(\mathbf{R}^+, X)$;
2. *the set* $(p_t^+ x;\ t \in \mathbf{R}^+)$ *is totally bounded in* $C_s(\mathbf{R}^+, X)$.

Before making use of this theorem in our further consideration, we give some auxiliary results.

Definition. *Let* d *be a function on* $\mathbf{R} \times \mathbf{R}$ *into* $\mathbf{R}$. *We shall say that* d *has* AP-*property iff to any sequence* $(t_n;\ n \in \mathbf{N})$ *there exists a subsequence* $(s_n;\ n \in \mathbf{N}) \subset (t_n;\ n \in \mathbf{N})$ *such that* $\lim_{n,m\to\infty} d(s_n, s_m) = 0$.

Lemma 3. *Let* X *be a Banach space,* $x \in C_s(\mathbf{R}^+, X)$ *and let for any* $t, s_1, s_2 \in \mathbf{R}^+$

$$(5) \qquad |x(t+s_1) - x(t+s_2)|_X^2 \le K\, e^{-at} |x(s_1) - x(s_2)|_X^2 + d(s_1, s_2) \quad ,$$

hold where $K, a > 0$ *are constants and* d *has* AP-*property. Then the set* $(p_t^+ x;\ t \in \mathbf{R}^+)$ *is totally bounded in* $C_s(\mathbf{R}^+, X)$.

In order to prove the existence of asymptotically almost periodic solutions of the equation (1) we shall find a solution of (1) which satisfies the estimate (5) and then we shall apply the theorem of Fréchet. We shall suppose that the functions f,g in the equation (1) satisfy the following assumptions:

(6)

a) f,g are almost periodic in t uniformly with respect to any ball in E ;

b) to any $r > 0$ there exists $k(r) > 0$ such that $\mathrm{Re}\,(f(t,x_1,y_1) - f(t,x_2,y_2),\ y_1 - y_2)_H \le -k(r)|y_1 - y_2|_H^2$ for any $t \in \mathbf{R}$ and $|[x_j, y_j]|_E \le r$ $(j=1,2)$;

c) to any $r > 0$ there exists $\ell_f(r)$ such that $|f(t,x_1,y_1) - f(t,x_2,y_2)|_H \le \ell_f(r)\,|y_1 - y_2|_H$ for any $t \in \mathbf{R}$ and $|[x_j, y_j]|_E \le r$ $(j=1,2)$;

d) to any $r > 0$ there exists $\ell_g(r)$ such that $|g(t,x_1,y_1) - g(t,x_2,y_2)|_H \le \ell_g(r)\,|u_1 - u_2|_E$ $(u_j = [x_j, y_j],\ j=1,2)$ for $t \in \mathbf{R}$ and $|[x_j, y_j]|_E \le r$ $(j=1,2)$.

The assumptions b) and c) are very restrictive (e.g., they imply that the function f may not depend on the second argument). The assumption a) says in other words that the functions $d_{f,r}$ and $d_{g,r}$, defined by

$$(7) \qquad \begin{aligned} d_{f,r}(s_1,s_2) &= \sup\,(|f(t+s_1,x,y) - f(t+s_2,x,y)|_H\ ;\ t \in \mathbf{R},\ |[x,y]|_E \le r) \\ d_{g,r}(s_1,s_2) &= \sup\,(|g(t+s_1,x,y) - g(t+s_2,x,y)|_H\ ;\ t \in \mathbf{R},\ |[x,y]|_E \le r)\ , \end{aligned}$$

have AP-property.

Let $u = [x,y] \in C_s(\mathbf{R}^+, E)$ be a solution of (1), $|u|_\infty \le r$. Further, let $s_1, s_2 \in \mathbf{R}^+$ and let us denote by $u_1 = [x_1, y_1]$ the function defined by $u_1(t) = u(t+s_1) - u(t+s_2)$. Then the function u_1 is a solution of the equation $x''(t) + Ax(t) = h(t)$, where

$$h(t) = f(t+s_1, x(t+s_1), y(t+s_1)) - f(t+s_2, x(t+s_2), y(t+s_2)) + $$
$$+ \varepsilon(g(t+s_1, x(t+s_1), y(t+s_1)) - g(t+s_2, x(t+s_2), y(t+s_2)))$$

and hence by Lemma 1 for any $t \in \mathbf{R}^+$ it holds

$$|u_1(t)|_E^2 + 2c\,\mathrm{Re}(x_1(t), y_1(t))_H = e^{-2ct}(|u_1(0)|_E^2 +$$
$$+ 2c\,\mathrm{Re}(x_1(0), y_1(0))_H) + 2\int_0^t e^{-2c(t-s)}\, h_1(s)\, ds ,$$

where $h_1(s) = \mathrm{Re}(h(s) + 2cy_1(s), y_1(s) + cx_1(s))_H$.

From the assumptions (6) it follows that for $c > 0$

$$h_1(s) \le (2c - k(r))|y_1(s)|_H^2 + c(\ell_f(r) + 2c)|y_1(s)|_H |x_1(s)|_H +$$
$$+ \varepsilon \ell_g(r)(1+ck)|u_1(s)|_E^2 + 2r(1+ck)(d_{f,r}(s_1,s_2) + \varepsilon d_{g,r}(s_1,s_2)).$$

Let now $c > 0$ and $\varepsilon_2 > 0$ be such that $|2c\mathrm{Re}(x,y)_H| \le 2^{-1}|[x,y]|_E^2$ for $[x,y] \in E$, $2c < k(r)$ and $\varepsilon_2 2(c^{-1}+k)\ell_g(r) + ck^2(\ell_f(r)+2c)^2(2(k(r)-2c))^{-1} = K_3 \in (0,1)$. Then we have for $\varepsilon \in \langle 0, \varepsilon_2\rangle$ and for $t \in \mathbf{R}^+$ the estimate

$$|u_1(t)|_E^2 \le 3e^{-2ct}|u_1(0)|_E^2 + 4r(c^{-1}+k)(d_{f,r}(s_1,s_2) +$$
$$+ \varepsilon_2 d_{g,r}(s_1,s_2)) + 2c\,K_3 \int_0^t e^{-2c(t-s)}|u_1(s)|_E^2\, ds .$$

Now we shall use the following simple

Lemma 4. *Let* z *be a continuous nonnegative function on* $\mathbf{R}^+$ *such that for some* $c > 0$, $K_1, K_2 \ge 0$ *and* $K_3 \in (0,1)$ *it holds for any* $t \in \mathbf{R}^+$

$$z(t) \le K_1\, e^{-2ct} + K_2 + 2c\,K_3 \int_0^t e^{-2c(t-s)}\, z(s)\, ds .$$

Then $z(t) \le K_1\, e^{2c(K_3-1)t} + K_2\,(1-K_3)^{-1}$ $(t \in \mathbf{R}^+)$.

By this Lemma it follows immediately from the above that for $\varepsilon \in \langle 0, \varepsilon_2 \rangle$ the function u satisfies the assumptions of Lemma 3 and hence $u \in AAP(\mathbf{R}^+, E)$. Hence we have proved the following

Theorem 2. *Let the equation (1) be given and let the assumptions (2), (4) and (6) be satisfied. Then to any* $r > 0$ *there exists* $\varepsilon_2 > 0$ *such that:*

a) *to any* $u_0 \in E$, $|u_0|_E \le r$ *and* $\varepsilon \in \langle 0, \varepsilon_2 \rangle$ *there exists a solution* u_ε *of the equation (1) on the interval* $\mathbf{R}^+$ *such that* $u_\varepsilon(0) = u_0$;

b) *if* u_ε *is a solution of the equation (1) on the interval* $\mathbf{R}^+$ *with* $|u_\varepsilon(0)|_E \le r$ *and* $\varepsilon \in \langle 0, \varepsilon_2 \rangle$, *then* $u_\varepsilon \in AAP(\mathbf{R}^+, E)$.

REFERENCES

[1] Fréchet M., Les fonctions asymptotiquement presque-periodiques continues. C.R. Acad. Sci. Paris, 213 (1941), 520-522

[2] Amerio L., Prouse G., Almost periodic functions and functional equations (Van Nostrand, New York 1971)

[3] Lovicar V., Almost periodic functions and almost periodic solutions of partial differential equations (Praha, 1973, in Czech, unpublished)

[4] Toušek M., Periodic solutions of partial differential equations of evolution type (Praha, 1973, in Czech, unpublished)

[5] Lovicar V., Toušek M., Bounded and asymptotically almost periodic solutions of the second order equations with nonlinear dissipative term (to appear)

Matematický ústav ČSAV, Žitná 25, 115 67 Praha 1, Czechoslovakia

A NEW TYPE OF GENERALIZED SOLUTION OF THE DIRICHLET PROBLEM FOR THE HEAT EQUATION

Jaroslav Lukeš
Praha (Czechoslovakia)

Let us consider in the Euclidean space $\mathbf{R}^n$ harmonic functions as continuous solutions of the Laplace differential equation $\Delta f = 0$. Given a bounded open set $U \subset \mathbf{R}^n$ and a continuous function f on the boundary U^* of U, we understand by the solution of the Dirichlet problem for f a continuous function F on the closure $\overline{U}$ of U which is harmonic in U and coincides with f on U^*. A set U is termed *regular* if there exists a solution of the Dirichlet problem for any continuous function f on U^* and, besides that, it is non-negative if f is. Of course, not every open bounded set in $\mathbf{R}^n$ is regular. There exist continuous functions on such sets for which we cannot solve the Dirichlet problem. Nevertheless, we can assign to those functions something like a solution in a reasonable way. If we denote for a continuous function f on U^* by H_f^U the infimum of all superharmonic functions on U whose limes inferior is at every boundary point z greater or equal to f(z), then H_f^U is a harmonic function on U and it is called a *generalized solution* of the Dirichlet problem for f obtained by the Perron method. Briefly, we shall call H_f^U the Perron solution of f. A point $z \in U$ is called a *regular boundary point* of U if $\lim_{x \to z} H_f^U(x) = f(z)$ for any continuous function f on U^*. The remaining points of U are termed *irregular*.

---

*) This paper is an expanded version of a communication submitted for publication in Comment. Math. Univ. Carolinae 14, 773-778 (1973).

We can construct the generalized solution H_f^U also by other methods, For instance, by Wiener method we extend the given function f onto the whole closure $\overline{U}$ and exhaust U by regular sets for which we can solve the Dirichlet problem, taking the limit of such solutions at any point of U. We obtain again the Perron solution H_f^U. This fact follows, for instance, from the following

Theorem (*M.V. KELDYCH* 1941, [3]). *Let* U *be an open bounded set in* $\mathbf{R}^n$ *and let* Φ *be a linear and monotone map associating with any continuous function* f *on* U^* *the harmonic function* $\Phi(f)$ *on* U *having the property that* $\Phi(F) = F$ *if* F *is continuous on* $\overline{U}$ *and harmonic in* U. *Then* $\Phi(f)$ *is equal to the Perron solution* H_f^U *for any continuous* f *on* U^*.

We mention still another method for the construction of the generalized solution H_f^U. If p is a potential and U is a bounded open set in $\mathbf{R}^n$, we denote by R_p^{CU} the infimum of all superharmonic functions which are greater than or equal to p on the complement CU of U. The greatest lower semi-continuous minorant of R_p^{CU} is called the *balayage* of p on CU and denoted by $\hat{R}_p^{CU}$. We may now formulate the following

Proposition. *For any* $x \in \overline{U}$ *there exists a unique Radon measure* μ_x^U *on* $\mathbf{R}^n$, *whose support is contained in* U^*, *such that* $\mu_x^U(p) = \hat{R}_p^{CU}(x)$ *for every potential* p.

Let us observe that $\mu_x^U(f) = H_f^U(x)$ for any continuous function f on U^* and for any $x \in U$. Further, a point $z \in U^*$ is regular if and only if μ_z^U is a Dirac measure at z.

According to the preceding proposition we may extend the definition of H_f^U by means of the balayage to the points of the boundary U^* of U. Therefore, for any $x \in \overline{U}$ we understand $H_f^U(x)$ to be a Radon measure on U^*. The following theorem is important.

Theorem. *Given a continuous function* f *on* U^*, *denote by* F *the restriction of* H_f^U *to* U^*. *Then* F *is a Borel function and* $H_F^U = H_f^U$ *on* $\overline{U}$.

The equality $\hat{R}^{CU}_{\hat{R}_p^{CU}} = \hat{R}_p^{CU}$ for any potential p is essential for the proof of this theorem. It follows easily from the facts that the set $\{x \in CU;\ \hat{R}_p^{CU}(x) < p(x)\}$ is polar and that $\hat{R}_p^{CU}$ is equal to the infimum of all superharmonic functions which are greater or equal to p on CU with the exception of a polar set.

Next we shall be interested in analogous problems for the heat equation. Let us consider now in the Euclidean space $\mathbf{R}^{n+1}$ "harmonic" functions - some authors use the term "parabolic" - as continuous solutions of the heat equation $\Delta f = \frac{\partial f}{\partial t}$. In the same manner as for the Laplace equation we define the generalized solution H_f^U of the Dirichlet problem by the Perron method or equivalently by means of the balayaged functions. The direct application of the Wiener method is not useful here since an exhausting by regular sets need not always exist. However, we may construct in this case more "generalized solutions". The Keldych theorem is no more valid for the heat equation. Let us mention the following example in $\mathbf{R}^2$. We set $U = (0,1) \times (0,1) \cup (0,1) \times (1,2)$, $V = (0,1) \times (0,2)$ and for any continuous function f on U^* we put $\Phi(f) = H_f^V$. It is easy to verify that Φ is a linear and monotone map and that $\Phi(f)$ coincides with the classical solution of the Dirichlet problem if this exists. But $\Phi(f)$ is not equal to H_f^U for every continuous function f. Likewise, the equality $\hat{R}^{CU}_{\hat{R}_p^{CU}} = \hat{R}_p^{CU}$ fails for the heat equation. This is caused by the fact that the role of polar sets of the Laplace equation is played by semi-polar sets, which need not be so "small". For the same reason also the equality $H_f^U = H_F^U$ fails, where F denotes the restriction of H_f^U to U^*.

The aim of this note is to introduce such a "principal balayage" T_p^{CU} for which the equality $\hat{R}^{CU}_{T_p^{CU}} = T_p^{CU}$ holds and by means of this principal balayage to derive a corresponding "principal solution" of the Dirichlet problem. Balayage determined in this manner is not unique, we need still a certain maximality condition.

The theory of harmonic functions derived from the Laplace equation or from the heat equation is a model for the general axiomatic theory of abstract harmonic spaces. In what follows we shall work in terms of this theory whose basic axioms, definitions and notation we should like to recall briefly. Let $(X, \mathscr{H})$ be a strong harmonic space in the sense of Bauer's axiomatic [1]. This means a locally compact Hausdorff space X with a countable base which is equipped with a sheaf $\mathscr{H}$ associating with any open set $U \subset X$ a vector space $\mathscr{H}(U)$ of real continuous functions, termed harmonic functions on U, such that the following axioms are satisfied.

Sheaf axiom. If $U \subset V$ are open sets and h is harmonic in V, then the restriction of h to U is harmonic in U and if $\{U_\alpha\}$ is a system of open sets and h is a function on $U = \bigcup_\alpha U_\alpha$ which is harmonic on any U_α, then h is harmonic on U.

Basis axiom. The regular sets (in the sense described above) form a base of X.

Convergence axiom. The limit of any increasing sequence of harmonic functions on any open set of X is a harmonic function whenever it is finite on a dense set.

Positivity axiom. For any point $x \in X$ there exists a harmonic function defined on a neighbourhood of x that does not vanish at x.

A function s defined on an open set U is termed *hyperharmonic*, if it is lower semi-continuous and if for any regular set V with $\overline{V} \subset U$ and for any continuous function f on V^* satisfying $f \leq s$ on V^*, the inequality $H_f^V \leq s$ holds on V (here H_f^V stands for the continuous extension of f to $\overline{V}$ which is harmonic in V). A hyperharmonic function finite on a dense subset of U is called *superharmonic*. A positive superharmonic function for which any positive harmonic minorant vanishes identically is called a *potential*. Now, the last axiom of our theory can be stated as follows.

Separation axiom. The set of all potentials on X separates the points of X, i.e., for any $x, y \in X$, $x \neq y$, there exist potentials p, q on X such that $p(x)q(y) \neq p(y)q(x)$.

The continuous solutions of the Laplace or the heat equations satisfy these axioms (in the case of the Laplace equation for $n = 2$ we must restrict the space to a bounded open subset of $\mathbf{R}^2$).

Let U be an open relatively compact subset of X. If p is a potential on X, then we define the *balayage* $\hat{R}_p^{CU}$ of p on CU as in the introduction. For every $x \in \bar{U}$ there exists a Radon measure μ_x^U, whose support lies in U^*, such that $\hat{R}_p^{CU}(x) = \mu_x^U(p)$ for any continuous potential p. We put further $H_f^U(x) = \mu_x^U(f)$ for any Borel function f on U^* and for any $x \in \bar{U}$. We may also obtain this generalized solution H_f^U of the Dirichlet problem on U by the Perron method and the μ_x^U for $x \in U$ is then called the *harmonic measure* on U at x. Since generally $\hat{R}_p^{CU}$ and $\hat{R}^{CU}_{R_p^{CU}}$ or - which is the same - H_f^U and H_F^U, where F denotes the restriction of H_f^U to U^* are not equal, we seek a new "balayage" and a new "generalized solution" without this defect. The construction is based on the following

Lemma. *Let* p *be a potential on* X. *If* $\mathcal{M}(p)$ *is the set of all potentials* q *such that* $q \leq p$ *and* $\hat{R}_q^{CU} = q$, *then the pointwise supremum of* $\mathcal{M}(p)$ *again belongs to it.*

We put then $T_p^{CU} = \sup \{q;\ q \text{ is a potential, } q \leq p,\ \hat{R}_q^{CU} = q\}$ for any potential p. We know that $\hat{R}^{CU}_{T_p^{CU}} = T_p^{CU}$. The potential T_p^{CU} is called the *principal balayage* of p on CU. Again, the next proposition is crucial.

Proposition. *For every* $x \in \bar{U}$ *there exists a uniquely determined Radon measure* ν_x^U *on* X, *whose support is contained in* U^*, *such that* $\nu_x^U(p) = T_p^{CU}(x)$ *for any continuous potential* p *on* X.

Hence, we may put $L_f^U(x) = \nu_x^U(f)$ for any continuous function f on U^* and for any $x \in \bar{U}$. The function L_f^U defined on $\bar{U}$ is termed the *principal solution* of the Dirichlet problem for f. The essential properties of the principal solution are stated in the following assertions.

Proposition. *The function* L_f^U *is harmonic on* U, *it is of the first Baire class on* $\bar{U}$ *and continuous on* $\bar{U}$ *in the fine topology (= the*

coarsest topology on X which is finer than the initial topology and in which any hyperharmonic function is continuous).

Theorem. *If F denotes the restriction of L_f^U to U^*, then $H_F^U = L_f^U$ on $\bar{U}$.*

Theorem. *If s is a continuous function on $\bar{U}$ which is superharmonic in U, then $L_s^U \leq H_s^U \leq s$ on $\bar{U}$. In particular, $L_h^U = h$ on $\bar{U}$ for any function h continuous on $\bar{U}$ and harmonic in U.*

We see that the map $f \mapsto L_f^U$ is linear and monotone, that function L_f^U is harmonic in U and that $L_h^U = h$, if h is continuous on $\bar{U}$ and harmonic in U. In the case of the Laplace equation it follows from the Keldych theorem that $L_f^U = H_f^U$. We are interested in whether this equality holds also in our general setting. First, we mention the following generalization of the Keldych theorem.

Theorem (*M. BRELOT 1960,* [2]). *Suppose that the harmonic space $(X, \mathscr{H})$ satisfies the following axiom D of domination:*

> *For any locally bounded potential p on X and any relatively compact open set V of X, H_p^V is the greatest harmonic minorant of p on U.*

Let Φ be a linear and monotone map which associates with any continuous function f on U^* a harmonic function $\Phi(f)$ on U. If $\Phi(h) = h$ for any function h which is continuous on $\bar{U}$ and harmonic in U, then $\Phi(f) = H_f^U$.

Let us observe that the classical solutions of the Laplace equation in $\mathbf{R}^n$ satisfy this axiom, while the solutions of the heat equation do not, and that the following Brelot convergence property of harmonic functions follows from the axiom D:

> The limit function of any increasing sequence of harmonic functions on any open connected set of X is a harmonic function whenever it is finite at a point.

If axiom D is fulfilled, then the set of all irregular points of U is polar and it is known that any polar set is of harmonic μ_x^U-measure zero for every $x \in U$. Now, our last theorem can be formulated as follows.

Theorem. *The following assertions are equivalent:*

(i) $H_f^U = L_f^U$ *for any continuous function* f ,

(ii) $\widehat{R}_p^{CU} = T_p^{CU}$ *for any continuous potential* p ,

(iii) *the set of all irregular points of* U *is of* μ_x^U*-measure zero for every* $x \in U$.

REFERENCES

[1] H. Bauer, Harmonische Räume und ihre Potentialtheorie, Lecture Notes in Mathematics 22, Springer Verlag, Berlin/New York, 1966

[2] M. Brelot, Sur un théorème du prolongement fonctionnel de Keldych concernant le problème de Dirichlet, J. Analyse Math. 8 (1960/61), 273-288

[3] M.V. Keldych. On the resolutivity and stability of the Dirichlet problem (russian), Uspechi Mat. Nauk 8 (1941), 172-231

[4] J. Lukeš, Théorème de Keldych dans la théorie axiomatique de Bauer des fonctions harmoniques, to appear in Czech. Math. J.

Matematicko-fyzikální fakulta Karlovy university, Sokolovská 83, 186 00 Praha 8, Czechoslovakia

SOME REMARKS ON DIRICHLET PROBLEM

Jiří Veselý
Praha (Czechoslovakia)

In the theory of harmonic spaces (see [1], [3]) the Laplace and the heat equations can be investigated simultaneously. We shall do some remarks on boundary value problems from this abstract point of view.

Let D be a regular region in the Euclidean space $\mathbf{R}^m$, $m \geq 2$. Given a function $f \in C(\partial D)$ (continuous on the boundary ∂D) we can find the solution of the Dirichlet problem for the region D and the boundary condition f. We can express it on D by means of the harmonic measure μ_x $(x \in D)$ in the form

$$(1) \quad H_f(x) = \int f d\mu_x .$$

To solve the Dirichlet problem for D, it would be sufficient to determine the harmonic measure relative to D. Unfortunately, it is not easy to do this even for very simple D. This "direct method" can be modified. We write formally

$$(2) \quad H_f(x) = \int f \frac{\partial \mu_x}{\partial \nu} d\nu = \int f . G_x d\nu$$

with a fixed measure ν and try to find the function G defined on $D \times \operatorname{spt}\nu$. For example we can choose $\nu = \sigma$, the area measure on ∂D, or $\nu = \mu_{x_0}$ for some fixed $x_0 \in D$. In case of the Laplace equation for a ball and $\nu = \sigma$ this leads to Poisson´s formula. For an interval $I \in \mathbf{R}^{m+1}$, $m \geq 3$, we obtain a similar formula for the heat equation (in spite of the fact that I is not regular), which is relatively very complicated (see [3], § 3.3). Other related information can be found, for example, in [7] for the Laplace equation and in [8] for the heat equation. In every case the boundary ∂D must be more or less smooth and this "direct method" leads to the Martin representation.

Another method, which is an old one and may be found in many textbooks, is "indirect". Provided G is an appropriate function we define the function u by

$$(3) \qquad u(x) = \int G_x d\mu$$

where μ is a signed measure. For the Laplace equation on $\mathbf{R}^m$, $m \geq 3$ we can choose for example

$$G_x(y) = \|x - y\|^{2-m} .$$

Under some assumptions the function u is continuous and harmonic outside spt μ. In connection with our problem we should like to find the measure μ for which u in (3) is harmonic on D and can be continuously extended from D to $\bar{D}$ in such a way that this extension coincides with the prescribed function $f \in C(\partial D)$.

If now G denotes the fundamental solution of the Laplace equation or the heat equation, the above mentioned idea leads us to the investigation of functions

$$(4) \qquad \begin{aligned} Vg &= \int_D g(y).G(x - y)\, d\sigma(y) , \\ Wg &= \int_D g(y).\text{grad}_y\, G(x - y)\circ\nu(y)\, d\sigma(y) , \end{aligned}$$

where $\nu(y)$ is the exterior normal to D at $y \in D$. These functions are usually called the single layer and the double layer potential, respectively.

Properties of these functions were studied very often and many investigations on this subject start by the phrase: "Let D be a region with sufficiently smooth boundary ∂D, ...". The notions of surface of bounded curvature, piece-wise smooth surface or Lyapunov surface are also connected with this problem. Regions with "irregular boundaries" were studied in [2] and in [9] - [11]. Most of the following results belong to J. KRÁL. The methods have their origin in some deep results obtained beyond the frame of the potential theory.

We shall start with the Dirichlet problem for the Laplace equation in $\mathbf{R}^m$, $m \geq 3$. Suppose that $D \subset \mathbf{R}^m$ is an open set with compact boundary. Let $\mathscr{D}$ be the space of all infinitely differentiable functions ψ on $\mathbf{R}^m$ with compact support $\operatorname{spt}\psi$. We write $\mathscr{D}(x) = \{\psi \in \mathscr{D}; x \notin \operatorname{spt}\psi\}$ and define for all $x \in \mathbf{R}^m$, $\psi \in \mathscr{D}(x)$

(5) $$W\psi(x) = \int_D \operatorname{grad}\psi(y) \circ \frac{y - x}{\|y - x\|^m}\, dy .$$

The notation is natural - in case of D with smooth boundary, $W\psi(x)$ reduces to the classical double layer potential. This fact can be also used for the extension of $W\psi(x)$ from $\mathscr{D}(x)$ to $\mathscr{D}$ provided $x \notin \partial D$. Thus $W\psi(x)$ may be considered to be a distribution over $\mathscr{D}$ with support in ∂D.

We shall study three problems:

(a) representability $W\psi(x)$ by means of a measure,

(b) extensibility of Wf continuously to $\bar{D}$,

(c) existence of a solution of the corresponding operator equation.

The problem (a) may be reformulated in the following way: when there exists a signed measure ν_x for which

(6) $$W\psi(x) = \int_{\mathbf{R}^m} \psi(y)\, d\nu_x(y)$$

holds for every $\psi \in \mathscr{D}(x)$? By calculus methods we obtain from (5)

(7) $$W\psi(x) = \int_{\Gamma} dH_{m-1}(\Theta) \int_{D(\Theta)} \partial_\varrho S_x\psi(\varrho,\Theta)\, d\varrho ,$$

where Γ is the unit sphere in $\mathbf{R}^m$, H_k is the k-dimensional Hausdorff measure, $D(\Theta) = \{\varrho > 0; x + \varrho\Theta \in D\}$, and

$$S_x\psi(\varrho,\Theta) = \psi(x + \varrho\Theta) .$$

For $x \in \mathbf{R}^m$, $(r,\Theta) \in (0,\infty\rangle \times \Gamma$ we denote

(8) $$\mathscr{S}_x = \mathscr{S}_x(r,\Theta) = \{y \in \mathbf{R}^m; y = x + \varrho\Theta, \varrho \in (0,r)\} .$$

We call now a point $a \in \mathscr{S}_x$ a hit of $\mathscr{S}_x$ on D provided both sets

$$\mathscr{S}_x \cap D \cap \Omega_\varrho(a) , \quad (\mathscr{S}_x - D) \cap \Omega_\varrho(a)$$

have positive H_1-measure for every

$$\Omega_\varrho(a) = \{y \in \mathbf{R}^m ; \|y - a\| < \varrho\} , \quad \varrho > 0 .$$

The number of all hits of $\mathscr{S}_x(r,\Theta)$ on D will be denoted by $n_r(x,\Theta)$ and we put

$$(9) \quad v_r(x) = \int_\Gamma n_r(x,\Theta) \, dH_{m-1}(\Theta) .$$

For $r = \infty$ we write simply $v_\infty = v$. It can be shown that

$$(10) \quad \sup \{W\psi(x) ; \psi \in \mathscr{D}(x)\} = v(x)$$

and from the integral representation of linear functionals we obtain:

the measure ν_x from (6) exists iff

$$(11) \quad v(x) < \infty .$$

Supposing (11) and writing f in (6) instead of ψ we can extend $W\psi(x)$ from $\mathscr{D}$ to $Wf(x)$ defined over $C(\partial D)$. Denoting $M = \{x \in \mathbf{R}^m ; v(x) < \infty\}$ we arrive at the following interesting fact: there are just two possibilities,

$$\text{either} \quad \overline{\mathbf{R}^m - M} = \mathbf{R}^m \quad \text{or} \quad \mathbf{R}^m - \partial D \subset M .$$

The latter case takes place iff the perimeter P(D) of D (see [4] or [10]) is finite. We shall suppose

$$(12) \quad P(D) < \infty .$$

The solution of (b) is the following:

Wf can be continuously extended from D to $\overline{D}$ for any $f \in C(\partial D)$ iff

$$(13) \quad \sup \{v(x) ; x \in \partial D\} < \infty .$$

Denoting by $\tilde{W}f$ the extension corresponding to Wf, we are interested in solving the equation

$$\tilde{W}g \mid \partial D = f$$

with a given $f \in C(\partial D)$. Writing I for the identity operator on $C(\partial D)$ we can rewrite this equation in the form

$$(14) \qquad (I - \overline{W})\, g = f \; .$$

We are able to find the distance of $\overline{W}$ from the set of all compact operators acting on $C(\partial D)$ and apply the Riesz - Schauder theory. The corresponding results (we are not going into details) are the following:

if the quantity

$$(15 \qquad \lim_{r \to 0_+} \; \sup_{x \in \partial D} \; v_r(x)$$

is sufficiently small, then (14) has exactly one solution for every $f \in C(\partial D)$. It should be mentioned here that the investigation of the nullspace of $(I - \overline{W})$ is not easy.

The method described above can be applied also to the heat equation. We shall write $z = [x,t] \in \mathbf{R}^{m+1}$, $m \geq 3$, where $x \in \mathbf{R}^m$, $t \in \mathbf{R}^1$. For an open set $D \subset \mathbf{R}^m$ with compact boundary and $T > 0$ we denote $E = D \times (0,T)$. The set E (generalized cylinder) is not regular for the heat equation. Because of the fact we choose the boundary condition f from the space $C_0(\partial E)$ of all continuous functions f on $\partial D \times \langle 0,T \rangle$, for which $f(\partial D \times \{0\}) = 0$. $\mathcal{D}$ has the same meaning as above (now in $\mathbf{R}^{m+1}$) and we introduce similarly $\mathcal{D}(z)$ for any $z \in \mathbf{R}^{m+1}$.

Now we write for $z \in \mathbf{R}^{m+1}$, $\psi \in \mathcal{D}(z)$

$$(16) \qquad W\psi(z) = - \int_E \left(\sum_{j=1}^{m} \frac{\partial G(z - w)}{\partial w_j} \frac{\partial \psi}{\partial w_j} + G(z - w) \frac{\partial \psi}{\partial w_{m+1}} \right) dw \quad .$$

This functional is an analogue of the double layer heat potential and we can again solve the problems (a) - (c).

Similarly to (7) we obtain

$$(17)\qquad W\psi(z) = 2^{m-1}\int_{\Gamma} dH_{m-1}(\Theta)\int_{0}^{\infty} e^{-\gamma}\gamma^{m/2-1}d\gamma \int_{E(\Theta)} \partial_{\varrho} S_z\psi(\varrho,\gamma,\Theta)d\varrho,$$

where $E(\Theta) = \{\varrho > 0;\ x + \varrho\Theta \in D\}\ (= D(\Theta))$,

$$S_z\psi(\varrho,\gamma,\Theta) = \psi(x + \varrho\Theta,\ t - \frac{\varrho^2}{4\gamma}).$$

For $z \in \mathbf{R}^{m+1}$, $(r,\gamma,\Theta) \in (0,\infty\rangle \times (0,\infty) \times \Gamma$ we denote

$$(18)\qquad \mathscr{S}_z = \mathscr{S}_z(r,\gamma,\Theta) = \left\{ w \in \mathbf{R}^{m+1};\ w = [y,u],\ y = x + \varrho\Theta,\ u = t - \frac{\varrho^2}{4\gamma} \right\}.$$

The notion of hits of $\mathscr{S}_z$ on E can be introduced similarly as above. Denoting by $n_r(z,\gamma,\Theta)$ the number of all hits of $\mathscr{S}_z(r,\gamma,\Theta)$ on E we can find some relations between $n_r(z,\gamma,\Theta)$ and $n_r(x,\Theta)$. After a long process we obtain solutions to (a), (b) and (c). Because of the special shape of E (a cylinder) those solutions are formally almost the same as above for the Laplace equation (for details and some results mentioned see [13]).

For other open sets E more general than cylinders we can also introduce the functional $W\psi(z)$ by (16). In (17) the set $E(\Theta)$ is then more complicated, i.e.,

$$E(\Theta) = \left\{ \varrho > 0\ ;\ \left[x + \varrho\Theta,\ t - \frac{\varrho^2}{4\gamma} \right] \varepsilon E \right\},$$

and for this case a more sophisticated variation $\tilde{v} = \tilde{v}_\infty$, namely,

$$(19)\qquad \tilde{v}(z) = \int_{\Gamma} dH_{m-1}(\Theta)\int_{0}^{\infty} e^{-\gamma}\ \gamma^{m/2-1}\ n_\infty(z,\gamma,\Theta)\ d\gamma$$

must be introduced.

The methods described show the "natural limits" of the applicability of the Fredholm method and give us some integral representation of the solution. The results obtained are very useful in connection with the second boundary value problem for the Laplace and the heat equations.

For sets with non-smooth boundary it is difficult even to formulate the above mentioned problem. We shall shortly describe the important steps (for the details see [10] and [11]):

First, generalized normal derivatives of single layer potentials are introduced. Under some "natural assumptions" they are represented by means of measures. Then the operators NV: $\mu \mapsto NV\mu$ of normal derivatives are introduced and one can solve the operator equations

(20) $NV\mu = \nu$

on the spaces of all signed measures. In case of the Laplace equation the dual operator to NV is the one we investigated in the beginning. In case of the heat equation the corresponding dual operator is very similar to that which was investigated above.

Some notions which we introduced and the methods described are good tools for investigation of angular limits (see [5], [13], [14]) and may be applied to the third (mixed) boundary value problems. For the Laplace equation this was done in [10].

REFERENCES

[1] H. Bauer, Harmonische Räume und ihre Potentialtheorie, Lecture Notes in Math. 22, Berlin 1966

[2] Ju.D. Burago, V.G. Mazja, V.P. Sapožnikova, K teorii potencialov dvojnogo i prostogo sloja dlja oblastej s nereguljarnymi granicami, Sbornik Problemy mat. analiza, Krajevyje zadaci i integralnyje uravnenija, Leningrad 1966

[3] C. Constantinescu, A. Cornea, Potential Theory on Harmonic Spaces, Berlin 1972

[4] E. deGiorgi, Nuovi teoremi relativi alle misure (r - 1)-dimensionali in uno Spazio ad r dimensioni, Ricerche Mat. 4, (1955), 95-113

[5] M. Dont, Non-tangential limits of the double layer potential, Čas. pro pěst. mat. 97, (1972), 231-238

[6] H. Federer, The Gauss-Green theorem, TAMS 58, (1945), 44-76

[7] R.A. Hunt, R.L. Wheeden, Positive Harmonic Functions on Lipschitz Domains, TAMS 147, (1970), 507-527

[8] J.T. Kemper, Temperatures in Several Variables: Kernel Functions, Representations, and Parabolic Boundary Values TAMS 167, (1972), 243-262

[9] J. Král, The Fredholm radius of an operator in potential theory I, II, Czech. Mat. J. 15 (90), 1965, 454-473, 565-588

[10] J. Král, The Fredholm Method in Potential Theory TAMS 125 (1966), 511-547

[11] J. Král, Flows of heat and the Fourier problem, Czech. Mat. J. 20 (95), 1970, 556-598

[12] I. Netuka, The third boundary value problem in potential theory, Czech. Mat. J. 22 (1972), 554-580

[13] J. Veselý, On the heat potential of the double distribution, Čas. pro pěst. mat. 98 (1973), 181-198

[14] J. Veselý, Úhlové limity potenciálů dvojvrstvy (English summary), Čas. pro pěst. mat. 95 (1970), 379-401

Matematicko-fyzikální fakulty Karlovy university, Malostranské n. 2/25, 110 00 Praha 1, Czechoslovakia

DIFFUSION PROCESSES AND THEIR CONNECTION TO PARTIAL DIFFERENTIAL EQUATIONS OF PARABOLIC TYPE

Ivo Vrkoč
Praha (Czechoslovakia)

The theory of Markovian processes and especially the theory of diffusion processes is very deeply connected with the theory of partial differential equations of parabolic type. The problem which is to be treated in the lecture originated in the theory of diffusion processes but can be reformulated and solved as a special problem from the theory of partial differential equations of parabolic type. For better understanding of both the problem and its modification it is necessary to begin with some facts from the theory of diffusion processes. It is also necessary to show some aspects of the relations between the two theories. For this reason the first part of the lecture is devoted to this matter. However, the necessary definitions and statements can be mentioned only very briefly and apart from its historical background.

First the well-known notions of diffusion and martingale processes will be needed.

Let a set Ω, a σ-field $\mathcal{F}$ of subsets of Ω and a nonnegative measure P which is defined on $\mathcal{F}$ and fulfils $P(\Omega) = 1$ be given. Suppose that a function $X(t,\omega) : \langle 0,\infty) \times \Omega \to \mathbf{R}^n$ ($\mathbf{R}^n$ is the n-dimensional Euclidean space) is defined almost everywhere (with respect to P) for every $t \in \langle 0,\infty)$. If, moreover, it is an $\mathcal{F}$-measurable mapping from Ω to $\mathbf{R}^n$ for all t we say that $X(t,\omega)$ is a stochastic process. Define $\mathcal{F}^t$ to be the smallest σ-field which is generated by $X(t,\omega)$ and $\mathcal{F}_t$ to be the smallest σ-field which is generated by $X(s,\omega)$, $0 \leqq s \leqq t$.

Definition 1. *A stochastic process* $X(t,\omega)$ *is called a Markov process if* $P(X(t,\omega)\in A/\mathcal{F}_s) = P(X(t,\omega)\in A/\mathcal{F}^s)$ *almost everywhere for every* s *and* $t, s \leqq t$ *and for every Borel set* A *from* $\mathbf{R}^n$.

The symbol $P(./\mathcal{F}_t)$ means the conditional probability with respect to the σ-field $\mathcal{F}_t$. It can be proved that there exists a modification of $P(X(t,\omega)/\mathcal{F}^s)$ such that a transition function can be defined $P(s,x;t,A) = P(X(t,\omega)\in A/\mathcal{F}^s)(\tilde{\omega})$ for $X(s,\tilde{\omega}) = x$. The transition function is measurable in s,x,t and it is a measure in A for fixed s,x,t. The transition function fulfils the well-known Kolmogorov - Fokker - Planck equation

$$P(s,x;t,A) = \int_{\mathbf{R}^n} P(s,x;r,dy)\, P(r,y;t,A) \text{ for } s \leqq r \leqq t .$$

Definition 2. *A Markov process is called a diffusion process if*

$$\text{i)} \quad \int_{|x-y|>\varepsilon} P(s,x;t,dy) = o(t-s) \textit{ for } s < t \textit{ and every } \varepsilon > 0$$

and if there exists a vector function $a(s,x)$ *and a matrix function* $b(s,x)$, *both defined on* $\langle 0,\infty)\times\mathbf{R}^n$, *so that*

$$\text{ii)} \quad \lim_{t\downarrow s} \frac{1}{t-s} \int_{|x-y|\leqq\varepsilon} (y-x)\, P(s,x;t,dy) = a(s,x) ,$$

$$\text{iii)} \quad \lim_{t\downarrow s} \frac{1}{t-s} \int_{|x-y|\leqq\varepsilon} (y-x)(y-x)^*\, P(s,x;t,dy) = b(s,x)$$

where z^* *is the transpose vector to the row-vector* z.

The vector function $a(s,x)$ *is called the drift of the process and the elements of the matrix* $b(s,x)$ *are called the diffusion coefficients.*

Under the condition that $a(s,x)$, $b(s,x)$ are Hölder continuous, bounded and $b(s,x)$ is uniformly positively definite and symmetric matrix there exists a diffusion process whose drift is $a(s,x)$ and whose matrix of diffusion coefficients is $b(s,x)$.

The diffusion processes are given by their transitions functions and by their initial states. Under the conditions formulated above the density $p(s,x;t,y)$ of the transition function $P(s,x;t,A)$

$(P(s,x;t,A) = \int_A p(s,x;t,y)dy)$ is the fundamental solution of the parabolic equation

$$\frac{\partial u}{\partial s} + \sum_i a_i(s,x)\,\frac{\partial u}{\partial x_i} + \frac{1}{2}\sum_{i,j} b_{ij}\,(s,x)\,\frac{\partial^2 u}{\partial x_i \partial x_j} = 0\ .$$

The method just described is one of the fundamental methods for constructing diffusion processes. Another method which was developed later consists in solving Ito stochastic equation

$$x(t) = x_0 + \int_0^t a(\tau,x(\tau))d\tau + \int_0^t c(\tau,x(\tau))\,dw(\tau)$$

where $w(t)$ is an n-dimensional Wiener process. This method is now described in many books on stochastic processes. See for example [1] ÷ [3]. To explain the method it would be necessary to define the Wiener integral and to establish its basic properties which is not possible in this short communication. By the standard method it is possible to construct the solution $X(t,\omega)$ under the conditions that $a(t,x)$ is a vector function and $c(t,x)$ is a matrix function continuous in t and Lipschitz continuous in x. In this case the drift of the solution $X(t,\omega)$ is $a(t,x)$ but the matrix of diffusion coefficients of $X(t,\omega)$ is given by $b(t,x) = c(t,x)c^*(t,x)$.

If we compare the assumptions of the method of Ito equations with the former one we see that the matrix of diffusion coefficients need not be positively definite but only positively semidefinite but $a(t,x)$,$b(t,x)$ are supposed to be Lipschitz continuous in x which is more restrictive than Hölder continuity needed in the former method. This gap between the assumptions of these two methods suggests that there could exist methods which should give us results under weaker conditions on positive definitness or smoothness of a,b. The most interesting articles which cover this gap and also use stochastic processes to define generalized solutions of parabolic equations are [4] to [6]. Even a rough explanation of the method requires the notion of martingale processes. Let an increasing family of σ-fields $\mathcal{F}_t$ (i.e. $\mathcal{F}_{t_1} \subset \mathcal{F}_{t_2}$ for $t_1 \leqq t_2$) be given such that $\mathcal{F}_t \subset \mathcal{F}$.

Definition 3. *A stochastic process* $X(t,\omega)$ *is a martingale with respect to* $\mathscr{F}_t$ *and* P *if*

i) $E(X(t,\omega)/\mathscr{F}_s)(\tilde{\omega}) = X(s,\tilde{\omega})$ *almost everywhere for all* $s \leqq t$,
ii) $X(t,\omega)$ is $\mathscr{F}_t$*-measurable for all* t.

$E(\,.\,/\mathscr{F}_s)$ is the conditional expectation with respect to $\mathscr{F}_s$. Equality i) is equivalent to $\int_\Lambda X(t,\omega)dP = \int_\Lambda X(s,\omega)dP$ for all $s \leqq t$ $\Lambda \in F_s$.

The problem of constructing a diffusion process to given drift and matrix of diffusion coefficients has been treated [4] in a special space Ω. The set Ω is the set of all continuous functions $\omega(t)$ on $\langle 0,\infty)$. The set Ω is equipped with the topology which is given by locally uniform convergence of $\omega_n(t) \to \omega(t)$. The σ-field $\mathscr{F}$ is the family of Borel sets in Ω. $\mathscr{F}_t$ are the smallest σ-fields generated by $\{\omega(s),\ 0 \leqq s \leqq t\}$. Instead of constructing directly a diffusion process, the authors solve the problem in terms of martingale processes. They introduce

Definition 4 [6]. *Let a vector function* $a(s,x)$ *and a matrix function* $b(s,x)$ *be given. A set of measures* $P_{s,x}$ *is a solution of the martingale problem if for every* $s \geqq 0$ *and every* $x \in \mathbf{R}^n$

i) $P_{s,x}\{\omega(.): \omega(s) = x\} = 1$,

ii) $f(\omega(t)-\int_s^t (L_\Theta f)(\omega(\Theta))d\Theta$ *is a* $P_{s,x}$ *martingale for every* $t \geqq s$,

$f \in C_0(R^n)$ *where* $L_\Theta = \sum_i a_i(\Theta,x)\frac{\partial}{\partial x_i} = \frac{1}{2}\sum_{i,j} b_{ij}(\Theta,x)\frac{\partial^2}{\partial x_i \partial x_j}$.

Theorem 1 [6]. *Let* $a(s,x)$ *be a vector function and* $b(s,x)$ *an* $n \times n$ *matrix function defined on* $\langle 0,\infty) \times \mathbf{R}^n$. *Assume that* $a(s,x)$ *is continuous and Lipschitz continuous in* x *and* $b \in C^{1,2}[\langle 0,\infty) \times \mathbf{R}^n]$ *is symmetric, positively semi-definite matrix. Then there exists a unique solution of the martingale problem.*

Moreover, there exists an extension $\tilde{\Omega}$ of Ω and extensions $\tilde{P}_{s,x}$ of measures $P_{s,x}$ and a $\tilde{P}_{s,x}$-Wiener process $\tilde{w}(t)$ so that

$$\omega(t) = x + \int_s^t a(\Theta,\omega(\Theta))d\Theta + \int_s^t b^{1/2}(\Theta,\omega(\Theta))d\tilde{w}(\Theta)$$

where the matrix $b^{1/2}(s,x)$ is a certain matrix fulfilling $b^{1/2}(s,x)(b^{1/2}(s,x))^* = b(s,x)$. This means that the process $X(t,\omega(.)) = \omega(t)$ is a solution of the given Ito stochastic equation and it is a diffusion process with the given drift and matrix of diffusion coefficients.

Let now u(s,x) be a solution of parabolic equation

$$(1) \quad \frac{\partial u}{\partial s} + \sum_{i=1}^{n} a_i(s,x)\frac{\partial u}{\partial x_i} + \frac{1}{2}\sum_{i,j=1}^{n} b_{ij}(s,x)\frac{\partial^2 u}{\partial x_i \partial x_j} = g(s,x)$$

Using Ito formula we obtain by simple calculation

$$Eu(t_2,\omega(t_2))/\mathcal{F}_{t_1}) = u(t_1,\omega(t_1)) +$$
$$+ \int_{t_1}^{t_2} E\left[\frac{\partial u}{\partial s} + \sum a_i \frac{\partial u}{\partial x_i} + \frac{1}{2}\sum b_{ij}\frac{\partial^2 u}{\partial x_i \partial x_j} /\mathcal{F}_{t_1}\right] d\Theta =$$
$$= u(t_1,\omega(t_1)) + \int_{t_1}^{t_2} E[g(r,\omega(r))/\mathcal{F}_r)]dr \quad \text{for } s \leqq t_1 \leqq t_2 .$$

It means that $u(t,\omega(t))$ is a $P_{s,x}$ martingale for every $s,x,t \geqq s$. This relation was used in [6] to generalize the notion of solution of parabolic equation. Let coefficients $a_i(s,x)$, $b_{ij}(s,x)$ and a function u(s,x) be given.

Definition 5 [*6*]. *The function* u(s,x) *is a generalized solution of (1) if*

$$(2) \quad u(t,\omega(t)) - \int_s^t g(\Theta,\omega(\Theta))\,d\Theta \text{ is a } P_{s,x} \text{ martingale for}$$

all $t \geqq s$ *and all* s,x *where* $P_{s,x}$ *is a solution of the martingale problem.*

Let the terminal condition u(T,x) = h(x) for the solution u be given. Using property (2) we obtain

$$u(s,x) = E_{s,x}u(s, \omega(s)) = E_{s,x}\Big[u(T, \omega(T)) - \int_s^T g(\Theta, \omega(\Theta))d\Theta\Big] =$$

$$= E_{s,x}h(\omega(T)) - \int_s^T E_{s,x}\, g(\Theta, \omega(\Theta))d\Theta$$

where $E_{s,x}$ stands for the conditional expectation

$$E_{s,x}\chi = E(\chi/\omega(s) = x) = \int \chi \, dP_{s,x} \, .$$

The previous formulas yield

$$(3) \qquad u(s,x) = E_{s,x}\, h(\omega(T)) - \int_s^T E_{s,x}\, g(\Theta, \omega(\Theta))d\Theta \, .$$

The explicit formula (3) shows that the existence of the solution of the martingale problem (i.e., the existence of measures $P_{s,x}$) implies the existence of generalized solution of parabolic equation (1) if only the functions h and g are integrable. Equality (3) gives us also the explanation of the meaning of the solution of parabolic equation in terms of stochastic processes. The conclusion can be formulated in

Theorem 2. *Let* $a(s,x)$, $b(s,x)$ *fulfil conditions of Theorem* 1 *and let* $h(x)$, $g(s,x)$ *be integrable functions. Then there exists a generalized solution of* (1) *fulfilling terminal condition* $u(T,x) = h(x)$.

The proof of the theorem follows immediately from the fact that $u(s,x)$ given by (3) fulfil condition (2).

The Dirichlet problem can be treated similarly. Assume that a region Q in n+1-dimensional Euclidean space $\mathbf{R}^{n+1}$ is given. Before defining the generalized solution of Dirichlet problem, we must first define the exit times. Put

$$\tau_s(\omega) = \inf\{t:[t, \omega(t)] \notin \bar{Q},\ t \geqq s\},$$

$$(4) \qquad \bar{Q} \text{ is the closure of } Q \, ,$$

$$\tilde{\tau}_s(\omega) = \inf\{t:[t, \omega(t)] \notin Q,\ t \geqq s\}.$$

Denote min (t_1, t_2) shortly by $t_1 \wedge t_2$.

Definition 6. *A function* $u(t,x)$ *is an outer generalized solution of (1) in the region* Q *if*

$$u(t \wedge \tau_s, \omega(t \wedge \tau_s)) - \int_s^{t \wedge \tau_s} g(\Theta, \omega(\Theta))d\Theta$$

is a $P_{s,x}$ *martingale for all* $t \geqq s$, $(s,x) \in \overline{Q}$.
A function $u(t,x)$ *is an inner generalized solution of (1) in the region* Q *if*

$$u(t \wedge \tau_s, \ (t \wedge \tau_s)) - \int_s^{t \wedge \tau_s} g(\Theta, \omega(\Theta))d\Theta$$

is a $P_{s,x}$ *martingale for all* $t \geqq s$, $(s,x) \in \overline{Q}$.
The measures $P_{s,x}$ *are the solution of the martingale problem corresponding to (1).*

Theorem 5.1 from [6] which is formulated for elliptic equations can be modified for parabolic equations.

Theorem 3. *Let* $a(s,x), b(s,x)$ *fulfil the conditions from Theorem 1, let* Q *be a cylindric region* $Q = (0,T) \times D$ *and* g, h *bounded measurable functions* $g: Q \to \mathbf{R}^1$, $h: (T \times D) \cup (\langle 0,T \rangle \times \partial D) \to \mathbf{R}^1$. *Then there exists a unique generalized outer solution* $u(t,x)$ *of* (1) *fulfilling the terminal condition* $u(T,x) = h(T,x)$ *for* $x \in D$ *and boundary condition* $u(t,x) = h(t,x)$ *on* Γ *where* Γ *is the set of points* (s,x) *for which* $P_{s,x}(\tau_s > s) = 0$ *(i.e.,* Γ *is a subset of* $\langle 0,T \rangle \times \partial D$).

The theorem can be proved by the same method as in [6], only the expression for the outer solution is now

$$u(s,x) = E_{s,x}\, h(T \wedge \tau_s, \omega(T \wedge \tau_s)) - \int_s^{T \wedge \tau_s} g(\Theta, \omega(\Theta))d\Theta \ .$$

An analogous theorem can be stated for the inner generalized solutions but the set Γ has to be replaced by a set with more complicated structure.

Varadhan - Stroock's characterization of outer solutions can be also modified for parabolic equations.

Let A be a Banach space of pairs of measurable functions (u,g) on $\overline{G}$. We say that (u_m, g_m) is star-converging to $(u,g) \in A$ if

$\lim u_m(s,x) = u(s,x)$, $\lim g_m(s,x) = g(s,x)$ for every $(s,x) \in \bar{G}$ and if $\sup |u_m(s,x)| > \infty$ and $\sup |g_m(s,x)| < \infty$. Let B be a set of pairs (u,g) such that u is sufficiently smooth and u is a classical solution of (1). Let $\bar{B}$ be the smallest closed set in A containing B with respect to the star convergence.

Theorem 4. *A function* $u(s,x)$ *is the outer generalized solution of* (1) *if and only if* $(u,g) \in \bar{B}$.

These definitions and results will be useful in the following problem. For the sake of simplicity we shall restrict ourselves to the one-dimensional case. Let D be an open interval $D = (x_1,x_2)$ and $Q = (0,T) \times D$. Let functions $a(t,x)$, $b(t,x)$ fulfil

(5) $a(t,x)$, $b(t,x)$ are defined on $\bar{Q}$;

(6) $a(t,x)$ is continuous and Lipschitz continuous in x;

(7) $b(t,x) \in C^{1,2}(\bar{Q})$, $b(t,x) \geqq 0$.

If (5) to (7) are satisfied then the assumptions of Theorem 1 are fulfilled which means that there exists a diffusion process with drift $a(t,x)$ and with diffusion coefficient $b(t,x)$. Let $\tau_s(\omega)$ be the first exit time defined by (4). Denote

$$P(b,x,a,Q) = P_{0,x}(\tau_0(\omega) \leqq T) \ , \quad x \in D \ .$$

The expression $P(b,x,a,Q)$ means the probability that the corresponding diffusion process (which starts at the moment 0 from the point x) leaves the closed interval $\bar{D}$ at least once within the time interval $\langle 0,T \rangle$.

Now let us assume that another diffusion process is given (which is described by a set of $\tilde{P}_{s,x}$) which has the same drift $a(t,x)$ but smaller diffusion coefficient $\tilde{b}(t,x)$, $\tilde{b}(t,x) \leqq b(t,x)$. This diffusion process gives us again the probability $P(\tilde{b},x,a,Q)$ (the region Q has not changed). The question is whether the probability $P(\tilde{b},x,a,Q)$ is smaller than $P(b,x,a,Q)$. Since generally the answer is negative (see examples in [7]) and this problem has some meaning in the theory of stability of randomly perturbated systems we give the precise definition. This definition differs from those given in [7], [8], [9] since $b(t,x)$ can be zero.

Definition 7. *Let* $a(t,x)$, $b(t,x)$ *be given in* $\overline{Q}, Q = (0,T) \times D$ *and fulfil conditions* (5) *to* (7). *The function* $b(t,x)$ *is maximal with respect to* $a(t,x)$ *and* Q *if*

$$P(b,x,a,Q) = \max_{b} P(b',x,a,Q)$$

for all $x \in D$ *where* $b'(t,x)$ *is any function defined on* $\overline{Q}$ *fulfilling* (5) *to* (7) *and* $b'(t,x) \leqq b(t,x)$ *in* Q.

We shall say that a function $q(x)$ *is convex locally at a point* x_0 *if there exists a neighbourhood of* x_0 *so that* $q(x)$ *is convex there in the usual sense.*

Now a theorem will be introduced which is the modification of Theorem 1 [7] to our case.

Theorem 5. *Let* $a(t,x)$, $b(t,x)$ *fulfil* (5) *to* (7) *in* $\overline{Q}$, $Q = (0,L) \times D$, D *is an open interval. The function* $b(t,x)$ *is maximal with respect to* $a(t,x)$ *and* Q *if and only if the outer generalized solution* $u(t,x)$ *of*

$$(8) \qquad \frac{\partial u}{\partial s} + a(s,x)\frac{\partial u}{\partial x} + \frac{1}{2} b(s,x) \frac{\partial^2 u}{\partial x^2} = 0$$

fulfilling $u(T,x) = 0$ *and* $u(t,x) = 1$ *on* Γ (*for definition of* Γ *see Theorem* 3) *is convex as a function of* x *locally at all points* $(t,x) \in Q$ *for which* $b(t,x) > 0$.

In the case that $b(t,x) > 0$ on $\overline{Q}$ the condition of Theorem 5 can be reformulated: $b(t,x)$ is maximal if and only if the bounded solution of (8) fulfilling $u(T,x) = 0$ and $u(t,x_1) = u(t,x_2) = 1$ for $t < T$ $(D = (x_1,x_2))$ is convex as a function of x in Q.

This theorem reduces the previous stochastic problem to a certain problem for parabolic equations. This new problem is to find conditions under which the given solution is a convex function.

Some conditions of this type can be given in the case that a and b do not depend on t.

Theorem 6 [8]. *Let functions* $a(x)$, $b(x)$ *fulfil conditions* (5) *to* (7) *in* $D = (x_1,x_2)$, $0 < K_1 \leqq b(x) \leqq K_2$ *and*

$$0 \leqq a(x) \leqq \frac{x_2-x}{2} \frac{K_2}{(x_2-x_1)^2} \arcsin^2 \sqrt{\frac{K_1}{K_2}} .$$

Then b(x) *is maximal with respect to* a(x) *and* Q, Q = (0,T) x D.

REFERENCES

[1] Dynkin E.B., Markovskie processy. Gosud. Izd. Fiz.-mat. lit., Moskva 1963

[2] Gichman I.I., Skorochod A.V., Vvedënije v teoriju slučajnych Processov. Izd. Nauka, Moskva 1965

[3] Gichman I.I., Skorochod A.V., Stochastičeskije differencial'nyje uravnënija. Izd. Naukova dumka, Kiev 1968

[4] Stroock D.W., Varadhan S.R.S., Diffusion processes with continuous coefficients I. Communications on Pure and Applied Mathematics, 1969, Vol. XXII, 345-400

[5] Stroock D.W., Varadhan R.S.R., Diffusion processes with continuous coefficients II. Communications on Pure and Applied Mathematics, 1969, Vol. XXII, 479-530

[6] Stroock D.W., Varadhan R.S.R., On Degenerate Elliptic-Parabolic Operators of Second Order and Their Associated Diffusions. Communications on Pure and Applied Mathematics, 1972, Vol. XXV, 651-713

[7] Vrkoč I., Some Maximum Principles for Stochastic Equations. Czech. Math. J., V. 19 (94) N 4, 569-604, 1969

[8] Vrkoč I., Some Explicit Conditions for Maximal Local Diffusions in One-Dimensional Case. Czech. Math. J., V. 21 (96), 236-256, 1971

[9] Vrkoč I., Conditions for Maximal Local Diffusions in Multi-Dimensional Case. Czech. Math. J., V. 22 (97), 393-422, 1972

Matematický ústav ČSAV, Žitná 25, 115 67 Praha 1, Czechoslovakia

SUBJECT INDEX

I

L

M

O

P

R

S

DATE DUE
